THE TREES AROUND YOU

HOW TO IDENTIFY COMMON NEIGHBORHOOD TREES IN THE PACIFIC NORTHWEST

CASEY CLAPP

MOUNTAINEERS
BOOKS

MOUNTAINEERS BOOKS is dedicated to the exploration, preservation, and enjoyment of outdoor and wilderness areas.

1001 SW Klickitat Way, Suite 201, Seattle, WA 98134
800-553-4453, www.mountaineersbooks.org

Printed in China

First edition, 2025

Design and layout: Jen Grable
All photographs and illustrations by the author except the following: *Siberian larch form* (p. 143), *black spruce* (pp. 99 and 106), and *Amur chokecherry form, bark, and flowers* (p. 313), all by Keith Turner; *Amur chokecherry leaf* (pp. 301 and 313) by Taha Ibrahimi
Cover photographs, front, clockwise from top: *Oregon white oak, bigleaf maple, sweetbay magnolia, fine gray fir cones*; back, top to bottom: *maritime pine, cornelian-cherry dogwood, Thundercloud plum, and holding up a head-sized sugar pine cone*; spine, *coast redwood*
Frontispiece: *Japanese katsura*

Library of Congress Cataloging-in-Publication Data is on file for this title.

Mountaineers Books titles may be purchased for corporate, educational, or other promotional sales, and our authors are available for a wide range of events. For information on special discounts or booking an author, contact our customer service at 800-553-4453 or mbooks@mountaineersbooks.org.

Printed on FSC®-certified materials

ISBN (paperback): 978-1-68051-744-6
ISBN (ebook): 978-1-68051-745-3

MIX
Paper | Supporting responsible forestry
FSC
www.fsc.org
FSC® C188448

An independent nonprofit publisher since 1960

To my mother, Michele Luffman, and my grandparents, Wayne and Barbara Rollinson and June and Carl Clapp—

Not only did you raise me, but you encouraged and supported me, even when my ambitions were outrageous and unorthodox. You let me explore, get dirty, and wax poetic, and you taught me strength, patience, precision, artistry, love, and perseverance. Everything good about me stands as a reflection of you.

CONTENTS

Broadleaf Trees 144

PREFACE

I love trees. I find them fascinating and beautiful—a constant source of curiosity and joy. I've spent many years studying them and working with them, trying to understand our relationship with them and how they function in this world. The more I have learned, the more I have come to understand that while collectively we know a lot about them, few people have a connection with trees on an individual basis, and I wondered why.

I wondered how people in my community—and in my culture more broadly—could simultaneously laud trees for their benefits and beauty but still hold them in such low priority compared to other aspects of culture. As I pondered these seemingly disparate perspectives, I was busy dipping my toes in the world of science education with my podcast *Completely Arbortrary*, as well as through countless classes and presentations on all sorts of topics relating to trees from an arborist's perspective. I found that a lot of people feel an intense desire to learn about trees, but for whatever reason, they find the topic inaccessible or intimidating. Sparingly few people know the names of the trees around them, or that there is such a massive diversity of trees—especially in our cities, as counterintuitive as that may sound. People tend either to not notice trees they do not recognize or to think they are all the same.

I wrote this guide to prove that sentiment wrong and to make getting to know the trees around you accessible. Indeed, that's why I do the work I do: I want to share how I see the world around me. I want people to really see and know their trees, to make their own connections to and develop their own relationships with their trees.

I live in Portland, Oregon, a big city filled with all the stuff you might expect: giant buildings, roads, houses, business districts, and cute neighborhoods. However, it's also filled with trees—a lot of them. Some are rare exotic species that you'll find only in a big city or a town park in our region; some are invasive species; others are relicts from our region's ancient past; some are small and unassuming but are among our oldest and most charismatic; some are simply stunning in their sheer enormity. If you know what you're looking at, a walk down your neighborhood street or to your local park is like walking through a botanic garden. I want to give everyone the power to see their neighborhood like this.

This tree guide is curated for the Pacific Northwest region and covers the trees that grow where most people live. It isn't a book about the trees in our wild forests, and it doesn't cover the numerous trees in the horticulture trade that we *could* plant here but generally don't. It is written for anyone to pick up this guide, take it outside, and identify the trees they find along their street, in their backyard, and in their local parks.

Of course, you can't learn it all in a day and it takes practice, but what doesn't? From riding a bike to playing *Pokémon*, we've all learned countless skills that required plenty of time and energy to master. Tree identification—any sort of plant identification really—is no different. I will show you that it's not as chal-

Casey looking at trees on a bed of Douglas-fir cones. (Warning: Do not attempt; you will get a lot of sap on your sweater.)

lenging as it may appear, but it certainly is rewarding.

It is my hope that picking up this guide changes how you see the world around you, that your eyes start drifting up a bit more, and instead of just noticing that there are some trees nearby, you begin to recognize them, like you would so many old friends. Learning the names of the trees around us is merely the beginning of rebuilding our relationships with these giant organisms that surround us almost everywhere we go.

With all of that said, this technical manual was written by me, and any and all errors or misrepresentations in it are mine and mine alone. Trying to capture a freeze-frame of the natural world is a fool's task, and I am as guilty as anyone else for attempting it; please forgive me where my efforts may have fallen short.

INTRODUCTION: WHAT IS A TREE?

There is no strict botanical definition of a tree. Trust me, I've spent far more hours debating this question than any self-respecting person probably should. Botanists call things that look like trees *arborescent*—literally "resembling a tree." Arborescence, though, is simply a form of growing, the life strategy of a plant, and many plants have achieved this form independently. To be a tree in the biological sense is to merely be treelike; it's very much a self-defining term, and any additional definitions are simply arbitrary, albeit perhaps useful.

This guide focuses on woody plants that tend to grow with a single stem (or potentially two or three) and attain a minimum height of about 15 feet (5 meters) at maturity. Excluded are plants that can sometimes become treelike but most often remain large shrubs. For instance, the common palm that is planted in the Pacific Northwest is included, but the bamboos are not. This approach mostly follows widely accepted convention among arborists, horticulturists, and dendrologists (those who study trees). Where

BELOW: *Each of these is called a tree, but what makes them a tree? On the left is a Canary Islands dragon tree, or* Dracaena draco, *and on the right is an ash tree, or* Fraxinus sp.

appropriate, closely related or commonly found look-alike nontrees are mentioned.

A **species** is a discrete population of individuals who are genetically similar to each other and distinct from other populations, and who all share a common **morphology**, meaning they look like one another, down to the last detail. One species of tree should be able to interbreed among itself but not with other species (though there are always exceptions, like hybrids). The term *species* is both plural and singular such that the statement "There are many species of conifers native to our region, and each species is unique" is correct.

A species is generally the lowest level of hierarchy in our system of **taxonomy**. Taxonomy is the scientific discipline concerned with naming, describing, and classifying life on Earth. It's meant to organize everything into more and more specific groups based on how closely they are related in an evolutionary sense. Plants that are more closely related to each other are grouped together lower on the taxonomic ladder (say, in the same genus or family), and those that are more distantly related are only grouped together at the highest levels (say, in the order or the class).

To see how taxonomic hierarchy is used, let's look at Scots pine.

Scots pine (Pinus sylvestris), an extremely common landscape species

KINGDOM: Plantae

DIVISION: Pinophyta

CLASS: Pinopsida

ORDER: Pinales

FAMILY: Pinaceae

GENUS: *Pinus*

SPECIES: *Pinus sylvestris*

Each species has a **scientific name**, often called a **binomial**, made up of a two-part Latin name: the **genus** and the **specific epithet**. Together they make a species name and refer to a single species. No two species have the same scientific name. In the example above, the genus is *Pinus* and the specific epithet is *sylvestris*, and together they name the species *Pinus sylvestris*—what we commonly call Scots pine. (Convention requires that the species name be written in italics to denote that it's a Latin word, and it's all lowercase except for the first letter in the genus, which is capitalized.)

The term genus (plural: *genera*) comes from the same Latin root as *generic* and *general* and refers to the second, broader level in the hierarchy. A genus includes all the species that are very closely related, like all the oaks (in the genus *Quercus*), the larches (in the genus *Larix*), or the birches (in the genus *Betula*). Closely related genera are nested within a **family**, and closely related families within an **order**, and so on up the ladder. An analogy for this is your surname or family name and your given name. My closest relatives and I all share the same family name, Clapp, but we each have our own given names. "The Clapps" refers to all of us (like the larches or the lindens), but Casey Clapp refers just to me.

This book focuses no higher than the family level and generally goes no lower than the species, though sometimes, where practical, different **varieties**, **subspecies**, and **cultivars** (a portmanteau of the words *cultivated* and *variety*) will be discussed. These lower rankings denote botanical or morphological differences between individuals or populations that taxonomists and horticulturists have determined warrant additional specificity. However, it is outside the scope of this guide to discuss these in depth; instead, a variety or cultivar will be noted only if it is important for figuring out the plant's identity. Several additional resources are listed at the end of the book that go further in depth on these more specific ranks.

There are several style conventions, notations, and abbreviations that are handy to know, as you'll see them throughout this guide. When referring

The curious giant sequoia cultivar 'Pendulum' (Sequoiadendron giganteum 'Pendulum')

to an unspecific species in a genus, the specific epithet is abbreviated as "sp." (e.g., "That is some kind of dogwood, *Cornus* sp."). When referring to multiple unspecific species in one genus, or to the genus broadly, the specific epithet is abbreviated as "spp." (e.g., "These are multiple different dogwoods, *Cornus* spp." or "All the crepemyrtles, *Lagerstroemia* spp., look the same"). A subspecies is denoted with "subsp." and a natural variety is denoted with "var." Note that the non-Latin terms and abbreviations are not italicized, but the actual names of the subspecies or varieties are (e.g., *Pinus contorta* subsp. *latifolia*).

Cultivar names add another level to our grand taxonomy that can be tricky to sort out. A cultivar is merely a named variety of a plant that has some unique trait that can be differentiated from the natural species and that is propagated by people. The name is given by whoever first found or propagated that variety, and it comes after the last italicized name of the natural species. When included in the scientific name, it's listed after the name in single quotation marks and is not set in italic (e.g., *Fraxinus angustifolia* subsp. *oxycarpa* 'Raywood'), or sometimes it follows the abbreviation "cv." Sometimes, when the parentage of the cultivar is far too complicated (mostly due to many hybridizations), the specific epithet is omitted entirely, such as in *Juglans* 'Paradox'. Though this is technically not a full name, it's often just easier, especially for the nursery trade. The first letter in each word of the cultivar name is capitalized.

The common name of a particular cultivar does not use quotation marks (e.g., *Fraxinus angustifolia* subsp. *oxycarpa* 'Raywood' would be called just Raywood ash). Finally, many cultivars have what's called a **trade name**, essentially a catchy name meant for advertising the cultivar—for example, Bountiful Blossoms magnolia or Orange Fury smoketree. These names can be the official cultivar name,

London planetree (Platanus x hispanica) is a very common hybrid in our landscapes.

but all too often the actual official name is something unsexy like 'Lmtsab2114' or 'Fastigiata', the first simply a reference code for the nursery, and the second an older, more scientific notation that has fallen out of fashion.

Hybrids also commonly occur in gardens and urban areas. Hybrids are the offspring of two parents that are different species. Examples include London planetree (*Platanus × hispanica*) and Leyland cypress (× *Hesperotropsis leylandii*). The "×" between the genus and the specific epithet denotes a hybrid origin from two species within the same genus, while an "×" before the genus indicates it's a hybrid from two species in different genera.

Trees all have **common names**—those most people use in everyday conversation, such as Oregon white oak, Japanese black pine, or western larch. For clarity and ease, this guide uses the generally accepted common name for a species in our area. However, there are significant regional differences between common names, even within the Pacific Northwest. Scientific names are helpful because they are accepted everywhere you go and refer to a single species. So, if you are in Seattle and mention Garry oak to someone from southern Oregon, they may not know that you are referring to what they call Oregon white oak. But if you say *Quercus garryana*, they will know exactly which tree you're referring to.

Common names can also be misleading from a botanical perspective—a perennial point of confusion when learning to identify trees. This is all too common with trees we call "cedars," such as the Port Orford-cedar or western redcedar. Botanically, these are in the cypress family, Cupressaceae, and are not closely related to the trees that we know botanically as the *true* cedars, which are in the genus *Cedrus* within the pine family, Pinaceae. These vernacular names were often given based on characteristics of the wood or

The large, upright cones of a true cedar (Cedrus sp.)

leaves or because they reminded Europeans of a tree they knew from Europe, but in most cases they do not reflect a tree's botanical identity. To help allay confusion, the names of trees that are botanically not what they are commonly called are properly written either with a hyphen between the words (e.g., Douglas-fir) or as one word (e.g., western redcedar). Conversely, the noble fir (a "true" fir in the genus *Abies*) has no dash and the Atlas cedar (a "true" cedar in the genus *Cedrus*) is written with a space.

Why Tree Identification Is Important

Everything about a tree begins with its name. For the professional, proper identification is the first step in working with trees. To assess a tree, whether its health, the risk of it falling, or the best management approach, you need to first know what kind of tree it is. For everyone else, it's the first step in connecting with and appreciating the giant, hulking organisms that tend to dominate our landscapes yet somehow also blend into the background.

Identification allows you to form a deeper relationship with trees, with implications far beyond a trivia game (and to be clear, it's great trivia). Like learning the names of your coworkers at your job, knowing the names of trees allows you to connect with them on a more intimate level. This personal connection strengthens your relationship to them and gives them value.

Plant awareness disparity (PAD, formerly *plant blindness*) is the phenomenon of people failing to notice the trees and plants around them. One result of this disparity is a shift in the general attitude in our society regarding plants (as opposed to societies that are more intrinsically plant-centric)—namely, that plants are less important than animals, including humans, and our built environments. PAD has been proposed as one reason for the lower level of support for plant conservation and preservation efforts: our modern societies simply do not value plants and trees as much as other aspects of the built world around us.

The problem with this phenomenon, ironically, is that plants are of paramount importance to us humans and all other animals, trees being of notable importance. Particularly in urban areas, trees are one of the most important parts of infrastructure, providing numerous health and climatic benefits. So, through identification, people develop awareness

Yoshino cherries (Prunus x yedoensis) along a street

Discussing the Oregon white oak on a tour

of the trees in their area, thus increasing the likelihood that they will support conservation efforts, whether in a natural area outside of town or in a construction zone down the street.

Eminently practical, sociologically important, and simply convenient, tree identification is a subtle yet important skill. Knowing the names of the massive, multiton organisms around us should be nearly as easy as naming the companies whose logos and brands we see every day, and in a real way, it can be. It turns out that we are extremely good at it.

Our brains are evolutionarily attuned to recognizing patterns in the natural world, and we're especially good at recognizing plants. It just so happens that today

we put those skills to work in other ways. Remember that *Pokémon* reference from the preface? I didn't choose it at random. A creative experiment in the UK found that young kids could identify, name, and classify different Pokémon with striking success, but they struggled to identify the common plants and animals that were actually around them. And it doesn't stop there: How many brands can the average American or Canadian name based on just a logo? How many cars could they name to a make or a model with just a quick glance? Probably more than they could the genus and species of their local trees.

Your brain is primed to recognize and name trees—it just needs a little direction.

HOW TO USE THIS GUIDE

Whether you work with trees every day or you're a recreational tree gazer, this guide is bound to be useful and practical. A few introductory sections, like this one, get you oriented, then it dives deep into the tree species, and finally concludes with some helpful information, including a glossary and additional resources.

As you may have already seen, the Introduction demystifies the naming of a tree in terms of its scientific name, how it's categorized, and its common name. In this section, you'll get a handle on how the book and tree profiles are organized.

The next section, The Stepwise Journey, covers two things: First, you'll get a breakdown of what you need to look at on a tree and what the various parts are called. It illuminates all the different characteristics that contribute to a tree's identity—or more specifically, what helps you figure out its identity. This is the most technical aspect of the book, but it's also the most helpful for getting your bearings if you're just starting out. Read through and refer back to that section whenever you need to; it's a tool to rely on until you're comfortable and confident enough to skip straight to the trees.

LEFT: *Two intrepid tree seekers identifying an Oregon white oak (Quercus garryana)*
RIGHT: *Leaves aren't required to identify the ominous figure of a black walnut (Juglans nigra).*

Following those essentials, you will learn the Stepwise Journey itself, my method for cohesively organizing all the abstract information in this book, as well as on the tree in front of you so that you can identify it. What should you look at first when trying to identify a tree? What comes second? What other clues should you consider? The Stepwise Journey is the best way I've found to narrow down the options for the tree I'm trying to identify. The section concludes with some handy lists and a calendar of flowering times to jump-start your tree identification journey. Think of them as cheat sheets to skip straight to the trees that have some notably unique trait that makes them stand out. (If you want even more in-depth information, flip to the back of the book for further resources on topics I only touch on, as well as a glossary to learn about any words you find confusing, many of which are set in blue.)

The heart of the book is all about the trees. It features detailed descriptions of the species that you're most likely to find where you live, and it's all about what you're looking for. I've broken down the main traits that set species apart with photos that help you know what to look for. Now, you may be tempted to skip straight to these profiles, but keep in mind that it's important to pin down the fundamentals first—they are the foundation upon which your skills will develop.

I've organized the tree profiles a bit differently than other guides. Namely, instead of by their evolutionary relationships, the trees are organized by how similar their leaves are. First, they are grouped by their leaf arrangement, then by whether their leaves are compound or not, and then by how similar their leaves or the whole trees look. I did this so that trees that may be more difficult to discern are closer together, and you can easily flip through a section and compare similar species. All the trees in one genus are kept together, so, for example, all the maples are next to each other, all the oaks are together, and so on.

When a genus contains multiple common trees, you'll find a general introduction to the genus to give you context on what sets it apart from other genera, where they are often found, and hints for identifying its species at the genus level (i.e., what makes an ash tree an ash). Information relevant to the whole genus can often be found here, but I've also included some helpful sidebars among species entries to help clarify important points. It's my hope that any questions you have while reading about a tree are addressed in these introductions or in the sidebars, but if you still have questions, refer to the resources in the back of the book, as one of those will surely address them.

Each species entry begins with a tree's common name, its scientific name, and a brief overview, including where it's most commonly found, both geographically and across the local landscape. If there are other commonly used vernacular names or old, outdated scientific names (called synonyms, referenced as "syn."), you'll also find them here. I've included the family name here or the introduction to the genus; family names always end with the suffix -aceae, as in Pinaceae, the pine family.

The name and broad overview are followed by a detailed description, including typical size, appearance, and key characteristics, and boldface headings call attention to bark, twigs, leaves, cones, flowers, and fruit, breaking up this information so that you can find what you need quickly when you're in the field. Specially important points about a particular part are set in italic. Entries conclude with "Species Remarks," some fun tidbits about the species, followed by descriptions of similar or related species and what to look for to tell them apart.

Given all that, you may be wondering how I choose which trees to include and

This landscape may look like a wild forest, but in fact it's in the middle of a city.

which to omit. Ultimately, it comes down to the probability that you will encounter them where you live. During the years I've spent as a professional arborist and dedicated tree seeker assessing and identifying trees across our region, it's become quite clear which trees are kept around in our towns and cities and which are relegated to wilder places. But the consideration of what to include goes further. I cross-referenced my own list of common trees with city tree inventories and planting lists from across our region, from southern Oregon to Vancouver, British Columbia, to Missoula, Boise, Seattle, Spokane, and Anchorage. I visited and surveyed trees in big cities and small towns and pored over horticulture books and landscape plant websites. I reviewed common tree lists from as many city foresters, arborists, and tree-focused nonprofits as would share them. Thus, I developed my final tally of common species based on which are being planted today, have been planted in the past, or that otherwise plant themselves.

Now, you will inevitably come across a tree that is not included in this book; we

grow dozens more species in our region than the few hundred I've covered here. But, in an effort to address unintended confusion in other tree books, I omitted those native species that are entirely absent from our cities and towns (or that are so rare, they might as well be absent) and those that make only occasional appearances, regardless of whether they are native or not. The reasoning is simple: you will have a greater chance of correctly identifying a tree if your options are narrower. It's far easier to contrast a few species than it is to compare ten or twenty. It doesn't matter how many species of linden in the world have fuzzy leaves if there's only one such tree in our region. So, when you find that one fuzzy-leafed linden, you'll automatically know exactly which one it is and how to differentiate it from the few other common lindens found here.

I've tried to cast as wide a net as possible to keep this guide useful. I mention if a species has a less common look-alike, or if there's a similar native species that may rarely sneak into the city. I've also included references for budding naturalists who want to take their training to the next level or find one of those rare trees that I've left out. If it's not in this guide, you can probably find it in one of the resources listed in the back of the book. But trust me, that tree you see outside the window right now? I bet it's in this book. The fun is figuring it out.

This Guide Is Just the Beginning

After you've waded through all the trees I've deemed common enough for the purposes of this guide, you will surely be wondering about those trees that were left out—and rightly so. Once you start to see trees that you recognize, you will inevitably start to notice those trees that you don't recognize, and that is where the real fun begins. This guide covers only about half of the region's native species, and there are probably one hundred more ornamental species that are simply not common enough to fit within this book's scope—not to mention the larger shrubs! This book is designed to serve as an introduction to the staggering diversity of woody plants in our region and to whet your appetite for seeking them out.

I encourage you to explore the resources listed in the back of the book. You can use them to get to know some of the lesser-known species and the rest of the region's native trees when you explore its wild forests. Some of the Pacific Northwest's most charismatic species don't leave the forest, and of those that do, some of the best examples can be found there. It's far easier to separate species in our native forests: there are far fewer of them, and they grow in predictable places and habitats; some can be identified based simply on where they are growing. You may also be pleasantly surprised when you find an uncommon ornamental on your street that isn't covered in this guide, spurring a deep dive into its identification. For a real challenge, head to your local arboretum or botanic garden to see the level of diversity that is possible in our region: the roughly 350 species covered in

Three budding naturalists are excited about identifying this curlleaf mountain-mahogany (Cercocarpus ledifolius).

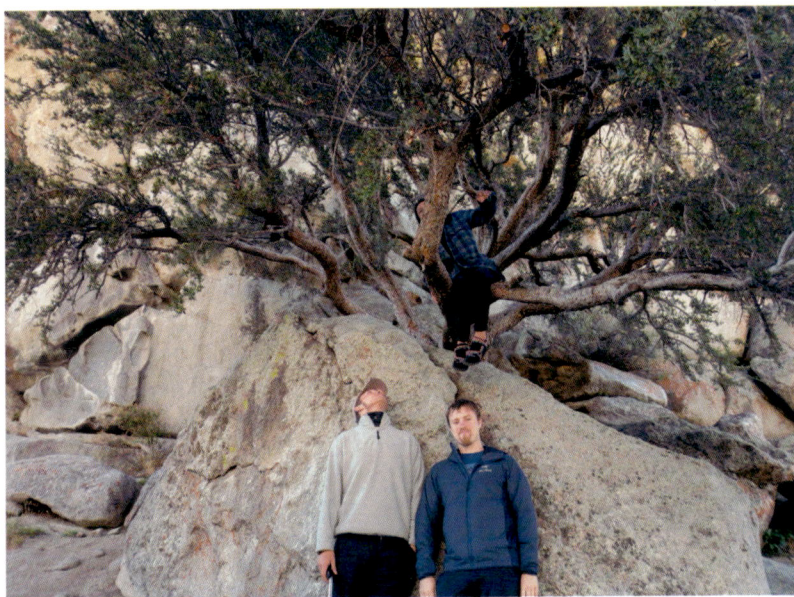

this guide are the tip of the iceberg when it comes to trees you may encounter.

Most of all, I encourage you to get outside and simply start looking at trees. I wrote this guide not only to make tree identification more accessible and show that anyone can do it but also to show off the amazing diversity in our cities and towns—the places where we live, work, and play. We walk under trees on our way to the store; we sit beneath them at pubs, restaurants, and coffee shops; they shade our back patios, houses, streets, and parks; and they play an outsize role in making where we live such a special and unique place. The trees in this book are the ones that have the greatest impact on our daily lives. They are also the ones at the greatest risk of disappearing and never coming back.

My hope is that this book helps you see the trees around you, notice how interesting and unique they are, and take action to help keep them around and to plant more. Every year our cities and towns lose more and more canopy, and all the while the warming climate continues to drive more extreme weather. Trees are our biggest assets in our existential struggle to deal with our negative impacts on our planet, and action is exactly what we need right now.

THE STEPWISE JOURNEY

When someone looks at a leaf and identifies the species, they are not simply looking at a random leaf and magically divining its provenance. They are essentially using muscle memory and performing several steps of an identification process while simultaneously weighing the options against a pool of candidates they know it may be—all in the span of a millisecond. Well, I guess it does sound a little magical, but that magic lives inside us all.

It boils down to the good ol' process of elimination. Many field guides feature what's called a dichotomous key, the botanical equivalent of Choose Your Own Adventure. You answer a series of questions about the tree, and as you choose between the options, the questions become more and more specific until it points you to a particular species. These kinds of keys can be very useful, but they often require specialized botanic knowledge, and the different parts of a plant they ask about may not be observable year-round. They also tend to be a bit reductive, in my opinion. So in this guide, I'd like to do things a bit differently by introducing the Stepwise Journey, a process of elimination I've used and taught for many years. Like a dichotomous key, it's a means of focusing your observations on specific characteristics when trying to identify a tree. What makes it different, and more intuitive, is that it encourages

LEFT: *What exactly am I looking at here?* RIGHT: *A hint that this is an Italian stone pine (Pinus pinea) is that it's in Italy.*

QUICK GUIDE TO THE STEPWISE JOURNEY

What should I look for first? What clues are most helpful? Examining the buds and leaves first will set you on the right path. Focus on one part at a time, compare what you see to what you discover in this guide.

Consider the context. Where is it growing? Was it planted intentionally, or does it seem to have self-seeded? Does the area look wild or well-maintained?

Look at form and texture. What shape is the tree? Is it evergreen? Upright? Weeping? Is the texture rough or sharp, soft or feathery? Are there many small limbs or fewer big ones?

Snag a twig and find the buds. Are they opposite or alternate? Is the twig thick or slender? Are the buds large or strikingly colored? Is the twig hairy or does it have growths or ridges? Are the buds clustered or on spur shoots? Are there thorns?

Define the leaf type. Are the leaves needle-like, scale-like, awl-like, or broad? Simple or compound? Palmately or pinnately veined and/or compound? (Narrowing down leaf arrangement and type will lead you to the correct section in this guide.)

Describe the leaves. For needle-leaf conifers, look at size, sharpness, shape, colors, and number of needles attached at a single point. For scale-leaf conifers, look at how the scales are arranged and the pattern on their undersides. For broadleaf trees, look at shape, size, edge, tip, and base.

Find the flowers. Flowers and fruit differ among genera and species. Are they symmetrical? What color are they? Are they arranged singly or grouped together, and in what formation? What time of year are they present?

Look for the fruit. Is it woody, papery, or juicy? What color and how big is it? Are they attached singly, connected, or grouped? What time of year does it appear?

Look at the bark. Describe the texture, color, and patterns. Are there any striking features like flakes, scales, or thorns? What color is it? Is it shaggy or hard, thick or smooth, uniformly patterned or irregularly geometrical?

Once you look closely at trees, you'll start to notice their differences, no matter how small. **Trust your eyes and don't overthink it**—when a tree matches the description and photographs featured in this guide, you've done it!

you to consider the whole tree and its context to help build your skill and proficiency as opposed to using only specific, narrow traits. It's not that reductive processes are bad; rather, a holistic approach can provide a richer and more accessible means of knowing a tree.

This straightforward, repeatable process of observation is more or less the same for any tree you want to identify. It begins with looking at the context of where and how the tree is growing, and then you focus on the most important

clues to quickly eliminate swaths of species that don't fit what you're seeing. Others only become important once you've narrowed your options a bit. Repeat this process for each tree, and you'll notice key differences between genera and species.

This guide already streamlines the pool of candidates to those you're most likely to find here in the Pacific Northwest and further organizes them by leaf arrangement, type, and shape—three important clues. The final steps are simply matching what you see to a particular group of trees, and then to an individual species by comparing the description with what you see.

An important aspect of learning about trees is learning how to describe them. Words help us visualize and focus our attention on distinct features. Once you can name a tree, you can more easily differentiate it, as your brain learns how to separate it from others. This also holds true for individual parts, like the shape of a leaf or the texture of bark. Applying words and descriptions to different characteristics helps you notice them.

If you're just starting out, simply go for a walk and look around to acquaint yourself with all the different facets of trees and their huge variety of forms without trying to identify them. This exercise trains your eyes to notice all the differences between trees, from the macro level to the tiniest detail. Of course, you'll need to learn a little about the species around you in order to go the final distance, and you'll need to memorize some things to become proficient. The more you practice this technique, the better you'll get, until it feels like you're magically divining a tree's name too.

Context

People who are proficient in tree identification use the context of where and how the tree is growing just as much as they use its attributes. The trees covered in this guide are the ones you'll see commonly in and around our cities and towns in the Pacific Northwest. But within these contextual bounds, there are further nuances and environmental gradients to consider (see Our Domestic Landscape, page 24).

Some spaces are heavily maintained and manicured—think your local park, a street planter strip, and most cemeteries. The trees in these areas are often intentionally planted and tend to be nonnative, ornamental shade trees. Conversely, some spaces are left to be colonized by whatever trees are tough enough to survive there—think steep, undeveloped hillsides, wetlands, and abandoned lots. Here, you'll tend to find a few adaptable **native species** or **nonnative**, **invasive species** that can successfully seed themselves in and compete with native trees. You are less likely to find noninvasive, **ornamental species** in these areas because they only grow where they have been intentionally planted.

CLEARING UP SOME KEY TERMS

Throughout this guide, I use terms with specific meanings that are often used flippantly in other contexts, especially on the internet, so I am setting the record straight. Here is how these terms are used in this guide:

- **native species:** A species that grew naturally in our region prior to the arrival and intervention of Europeans.
- **nonnative species:** A species that was not found in our region prior to the arrival and intervention of Europeans. A nonnative tree is not necessarily invasive or bad.
- **ornamental species:** A species that is intentionally planted as part of a maintained landscape. These are most often, but not necessarily, nonnative.
- **invasive species:** A nonnative species that has naturalized into the local environment and will readily grow from seed without any help from people, often competing with native species.

There are also gradients throughout our human landscapes, such as lowland to highland and rural to urban. Lowland spaces have more water-dependent or water-tolerant species, while upland spaces tend to be colonized by species that prefer drier conditions. In terms of urbanization, the largest developed cities tend to have higher species diversity, where nonnative species make up a larger proportion of species overall. As you move away from these population centers, the species diversity shrinks, and a greater proportion are native, often relicts of prior land use.

Indeed, you'll begin to notice that certain trees are planted almost everywhere under certain circumstances, like red maples in new suburban developments, or cherries and Thundercloud plums in front yards. Japanese maples are almost always showcased near front windows, and dogwoods are often planted where a Douglas-fir has been removed. Land use and regulation, or lack of it, can play a big role in what trees are sold where, not to mention the nursery industrial complex's role in what trees are planted (you'll find a correlation between certain trees and big-box stores).

Noticing these landscape patterns can help you know which species you're likely to encounter. As always, there are exceptions, like natural parks within big cities or unique plantings by inspired small towns. But those exceptions often prove the rule, and learning the patterns around your area will help immensely.

After thinking about where, notice how it's growing. Observe the whole tree, how its branches and stem are oriented, the texture of the twigs and foliage, the shape of the crown, and whether it has any fruit or flowers. They not only help define that species' unique character but also are a unique part of that individual tree based on where it's growing and what's around it. Trees adapt to their environments as they grow.

Our Domestic Landscape: Some Helpful Descriptions

Throughout this guide, I refer to different landscape types or areas within the landscape that different trees may be found growing in. These are my interpretations of the landscape; other people may not see them the same way. They are my working definitions of the landscape types that I see around our built environments, sort of urban ecological niches in our domesticated environments. They are meant as frames of reference, as no landscape is exactly like any other; however, I bet you'll be able to spot them in your area, too.

Yards

The first landscape is the private yard where homeowners control what is planted, kept, or removed; there are generally two kinds.

- **Older houses and neighborhoods built before the 1960s or 1970s,** where the landscape has had some time to develop and plant fashions have come and gone. Interesting trees have been planted here and there for decades because, at one time, big curious trees were all the rage. These usually include only the largest native species and a varied selection of ornamentals.

- **Newer houses and neighborhoods built after the 1970s,** where you find trees from larger nurseries or big-box stores that sell the most profitable and easy-to-grow species. These often include large, nonnative trees common across the country, a few native trees that have snuck in, and several invasive species. More often you will see a consistent selection of smaller ornamental trees and large shrubs.

Both yard types are common throughout our region, from rural towns to large cities. Houses with yards teeming with varied and interesting species are often the exception.

LEFT: *Newer developments often all have similar trees.* **RIGHT:** *Lavalle hawthorns are most commonly found as street trees rather than in yards.*

Red Maple Suburbs

The name gives it away: these kinds of housing developments usually maximize house footprints and minimize landscaping, including street trees. They have green lawns, identical houses, and the same exact trees in every yard or planter strip (if there are any at all). Trees planted here are often cheap and purchased in bulk, such as red maples, flowering cherries, zelkovas, and Callery pears. Other common trees include small-growing Colorado blue spruce, Eddie's White Wonder dogwood, and Leyland cypress, considered a "big tree" in a yard.

Professional Landscapes

The landscape of the ubiquitous professional business park has cheap, square buildings that take up less area than the parking lots around them. Trees here must be tough because quick-visiting landscape services blow away all the organic matter that lands in their tiny planting spaces before it can even start composting. These trees and plants are the who's who of tried-and-true (that is, cheap and durable) landscaping stock that meets the bare minimum requirements for landscaping codes.

Like those in the red maple suburbs, trees here generally all look the same. Here you'll find Scots pine, cherries, Nor-way and red maples, sweetgum, junipers, and green ash—and another half dozen species that are usually half-dead because they couldn't quite withstand the abuse.

Parks

These are the municipal pleasure grounds where people can take a rest or kids can play. Trees here can be a bit deceiving. Most of the time, you'll find classic, large-growing, ornamental shade trees like American elm, Norway maple, sugar maple, red oak, pin oak, and London planetree; or you may see popular native trees, such as Oregon white oak, Douglas-fir, and ponderosa pine, depending on where you are. However, parks also may include strange experimental trees—inspired choices that are a little more expensive or curious or that have some kind of importance—intermixed with the more mundane ones to keep things interesting.

Streets

This category refers specifically to the planter strips between the curb and sidewalk, or maybe in the middle of a roundabout or median—trees that are generally called "street trees." In some places, they are both required and regulated so that only appropriate trees are planted, while in other places, they are optional. In our largest cities, this is where you will find

Trees with invasive tendencies are opportunists, growing in less-than-optimal places, like this European white birch (left) or this Siberian elm (top right) growing in cracks in the pavement. **RIGHT, BOTTOM:** *Wild or unmanaged forested areas often all have similar trees.*

the highest diversity of species, including more curious and interesting trees than in any other landscape outside a specialty park or garden. Farther away from city centers, these spaces become more homogenized to a few species (as in red maple suburbs).

Roadsides

This landscape is that often-but-not-always maintained area between a road (usually without a sidewalk or curb) and the adjacent private property, a real no-man's-land of botanical wonder. Here, two sorts of trees dominate: native edge species that can finally grow in peace (the ones that can handle some pollution), such as Pacific dogwood, bigleaf maple, Oregon ash, and cottonwoods, and non-native, naturalized species that can also finally grow in peace, such as European white birch, English holly, English hawthorn, tree of heaven, sweet cherry, and Siberian elm. Farther outside larger cities, this area also tends to include the toughest species, such as western juniper, alders, and lodgepole pine.

Landscape Margins

The margins are the part of the landscape that is not quite maintained but also isn't exactly an unmanaged wild area. Look for these areas next to a fence line or behind a hedge, on the other side of the shed where no one goes or behind the retaining wall of the house next door, or in an empty lot or alleyway. This is the landscape for misfit trees. The soil is often poor, hot, and sparse, and trees grow between pavers and rock walls or at the base of debris piles. They must make their own way and curve toward the light—they can't depend on anyone but themselves for survival. This is where only the most adaptable native species, like bigleaf maple or cottonwood, grow. You'll also find the not-quite-invasive trees that get planted by squirrels, like red oak, pin oak, and English walnut, but also tougher invasives, like Norway and sycamore maples, English hawthorn, sweet cherry, European white birch, and English holly.

Unmanaged "Natural Areas"

These spaces are similar to the landscape margins, but instead of being around the edges of an otherwise managed landscape, they *are* the landscape. These are the undeveloped and unmanaged pockets around town, often of public or unknown ownership, left to go feral. Urban and suburban kids may remember playing in what felt like a wild forest in such places. Here, you'll find some native species that miraculously avoided the saw, such as Oregon white oak, Douglas-fir, ponderosa pine, and grand or white firs. But more often you'll find younger groves of adaptable native species—like bigleaf maple, red alder, cottonwoods, shore or lodgepole pine, and junipers—competing on equal terms with large local invasive species— like Siberian elm, Norway maple, English holly, and European white birch. This is a true forest but is composed of a melting pot of species.

Natural Area Margins

These areas are where the less disturbed, often native forest landscape bumps up to backyards: the green spaces left as part of a development on the edge of town (usually where there's a creek or drainage of some kind) or a neighboring property covered in native forest trees. It's here that a mix of the most common native species is found sneaking into backyards, along with a bevy of the most common but smaller invasives. In the eastern portion of our region, you'll often find western juniper, Russian-olive, Siberian elm, and a handful of shrubs. In the west, you're likely to see red alder, bigleaf maple, Douglas-fir, and Oregon ash, as well as invasive species like English hawthorn, sweet cherry, and Norway maple. Anything farther outside this area falls within the realm of books focused exclusively on the native trees of our region.

Form and Texture

"You can tell it's a fir tree because of the way it is." Well, yeah, you can. Knowing the way a tree *is*—aspects of its form, habit, texture, and color—is one of the most useful qualities, allowing you to pick out a particular species without even getting near it. Together, these traits combine to give a tree its own unique look, recognizable at a distance, or even from the seat of a moving vehicle.

The general shape of a tree is its **form**; form broadly refers to the overall appearance of the canopy and the architecture of the branches holding it up. It is heavily influenced by a tree's natural **habit** and its surrounding environment, such as whether it is growing out in the open or in a forest among other trees. When grown in the open, as in a front yard or along a street, trees of the same species develop the same archetypal habit and texture. Some habits are quite distinctive and immediately set a tree apart, while others can appear fairly generic. Either way, its habit and form provide information about a tree.

We split growth form into two generic categories: excurrent and decurrent. An **excurrent** growth habit is characterized by one central leader with horizontal branching all the way up the stem. The oldest branches are the lowest and longest, and as you go up the stem, the branches are younger and therefore shorter, creating a tapered appearance. This is the quintessential conifer shape— or as a kid would draw it, a triangle on a stick.

A **decurrent** form is characterized by one or more central leaders that grow up and out, with multiple larger limbs that branch out from leaders and also grow up and out. This is the quintessential growth habit of broadleaf trees, and we call the large, outstretching branches **scaffold limbs**. Scaffold limbs aren't quite main trunks; they are those large, outward-growing limbs that support

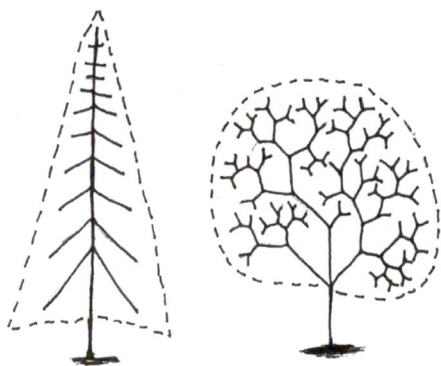

LEFT: *Excurrent form* RIGHT: *Decurrent form*

long and straight, while others may be short and curvy; some arc up and outward, and others snake horizontally or hang down. These somewhat subjective differences can be difficult to notice on a single tree, but they are clear as day when you look at two different species growing right next to each other.

The form and architecture of a tree are often expressed as its **silhouette**, and it can showcase the textural appearance of a tree very effectively, especially in winter. A tree's silhouette is determined by traits like twig size, density, and orientation as well as branch angle and limb curvature. This is equally true for broadleaf trees and conifers, except that silhouettes for most conifers include the texture of the evergreen foliage and how it's held on the twigs.

A tree's age and surroundings are factors to consider when looking at its form and architecture. Most young trees start with one dominant main leader and develop their mature form based on their surroundings. The **archetypal forms** for most trees are based on what they would look like growing out in the open and

significant amounts of the canopy. If a kid drew a decurrent tree, it would be a circle on a stick.

Within these two generic forms, you can discern further detail. Some excurrent trees are wider or skinnier, while others may be irregularly branched or uniformly triangular. Decurrent trees may be wider than they are tall or taller than wide, oval shaped or globe shaped, vaselike or irregularly chaotic. These more nuanced forms are often a function of the tree's branch architecture.

Branch architecture is how a tree looks as it holds itself up. Different trees have different-sized limbs attached at different angles. Some of those limbs will be

YOU CAN GROW YOUR OWN WAY

Decurrent growth is most often seen in broadleaf trees, but different species create it in their own ways. Some trees (like sugar maple, p. 173) maintain a central leader with many upward- and outward-growing limbs that create a very uniform, oval-shaped canopy. Others develop multiple main leaders that each support a huge chunk of the canopy and collectively form one big crown (like bigleaf maple, p. 181), or quickly diverge into many smaller, upward-growing branches that create a bouquet-like collection of stems (like Persian parrotia, p. 221).

Still others embrace anarchy and have no leaders; trees with this form are dominated by scaffold limbs, every branch for itself growing up and out (like Norway maple, p. 172, deciduous magnolias, p. 264, and most smaller broadleaf trees).

unaffected by nearby trees. We call these **open-grown** trees, and they develop their signature form based mostly on the fact that they can grow up and out to their heart's desire.

Near other trees or perhaps large buildings, trees grow up to where the light is and out until they run out of room or get too shaded. They tend to retain their dominant main leader to grow up, shedding their lower limbs that get shaded out. This is the **forest-grown** form, which isn't exactly accurate outside of a forest. However, the effect is the same: a tree will grow tall and skinny regardless of whether it's seeking light in a crowded forest canopy or from between two closely set apartment buildings.

The overall form of a tree is affected in significant ways by the cumulative effects of the tiniest details. The size and shape of the leaves, their arrangement and orientation, and the twigs they are attached to may be almost impossible to see when

Winter silhouettes can be strikingly different depending on the species; clockwise from top: European hornbeam, American elm, black walnut, European beech.

LEFT: *Trees that grew up in forests tend to be tall and skinny.* **RIGHT:** *Trees that grew up without other trees around are shorter and wider. These are both the same species, ponderosa pine.*

you look at a tree from a distance, but the patterns and textural appearance they produce across the whole canopy can be very apparent.

Texture is most often related to traits of the leaves, ranging from size to orientation to shade tolerance. For example, the bigger the leaves, the wider they are spaced to accommodate their neighbors without shading them too much. Smaller leaves are often packed more tightly together. Leaf size and spacing can make a canopy look rougher or more delicate. If the species is more tolerant of shade, then the canopy may have many overlapping leaves, creating a layered, dense appearance. Shade-intolerant species may have very few overlapping leaves and almost no leaves in the interior of the canopy.

There really is no limit to how a tree's attributes can influence its texture nor to the fun terms you can choose from to describe them: trees can be rough, regal, relaxed, lanky, layered, lazy, spiky, shaggy, sharp, soft, starlike, irregular, unkempt, airy, feathery, curvy, tufty, knobby, delicate, dense, and more. The important thing is that you try to describe

it and contrast it to similar trees to help you notice the small differences. Even the tiniest details can help.

A tree doesn't need leaves for you to make inferences based on texture; in the winter, twigs can be just as useful. Twig size and spacing tend to be roughly correlated with leaf size too: big leaves require proportionally stout twigs and a wider-spaced canopy, while small leaves can be supported by slender, more closely spaced twigs. The size and density of the twigs, their orientation, and how curvy or straight they are all combine to give a deciduous tree its unique texture in wintertime, when it often seems like a deciduous tree's form and texture blur together such that it's difficult to describe one without describing the other.

Regarding texture, conifers are no different from broadleaf trees: the attributes of their foliage often create their unique texture. The long, bunched needles of pines that grow outward from all around the twigs create a very different appearance from the smaller, stiff needles of a spruce; the feathery, relaxed texture of hemlocks contrasts with the rougher,

LEFT: *Color and texture help set apart the western larch (left) and a true fir (right).*
RIGHT: *The winter silhouette of a row of northern red oaks*

unkempt texture of Douglas-fir; and the rigid, compact true firs look very different from the pillowy false cedars. These textural differences combined with growth form can be enough to narrow down a tree to at least a genus—and often a species once you're familiar with the trees around you.

Following closely on the heels of texture is color. Contrary to popular belief, not all trees are green, and even the ones that are can differ markedly in hue and shade. The human eye can pick out different shades of green better than any other color, so the slightest differences can be important, especially for conifers.

And of course, don't forget about seasonal colors. The colors a tree turns in the fall and exactly when it changes can be very good clues as to which species it is. Some trees change very early, like Kentucky coffeetree, while others may not change at all and simply drop green leaves, like red alder. It can also be helpful to know whether a tree consistently displays a certain color that differs from other closely related or look-alike species. For example, Norway maple and sugar maple have similar leaves and can be tough to tell apart. However, their fall

colors differ markedly; Norway maple has a yellow, russet-colored display compared with sugar maple's fiery orange.

Twigs and Buds

Once you've got a feel for the whole tree, your next step is to look more closely at specific parts. Keep in mind that the form and texture are often influenced by these smaller parts, and ideally all these aspects will, through the Stepwise Journey, merge to inform your identification. The first part to focus on is the twig, or more specifically the buds.

Twigs are the most recent one to three years of growth at the end of a branch. They have not developed the mature patterns, colors, and textures of older limbs and thus can have unique traits that disappear with age. Twigs are important for two reasons: First, they hold the buds, which tell us important information about the leaves. Second, they are always present, so when the leaves of deciduous trees fall away, the twigs are the next best identification trait. The good news is twigs have a lot of little details to notice.

The physical attributes of the twig itself are a good place to start. Some twigs are stout, with diameters greater than ¾ inch

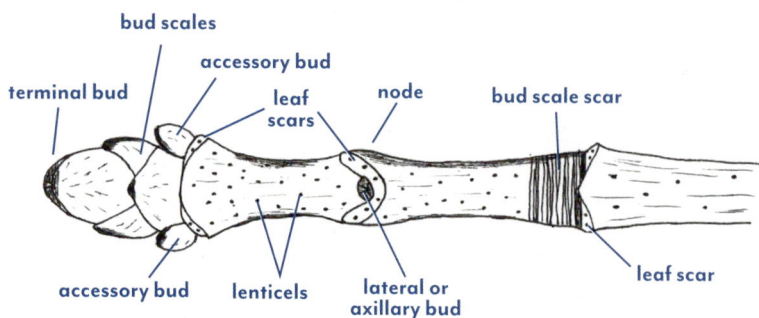

bud scales
accessory bud
terminal bud
leaf scars
node
bud scale scar
accessory bud
lenticels
lateral or axillary bud
leaf scar

Parts of a twig

(2 centimeters), while others are very slender (recall that this tends to correlate with leaf size). Some twigs are perfectly round in cross section, while others are angled. Twigs may zigzag or arch upward, creating a distinct pattern, and many have distinct colors ranging from red to green or brown to yellow.

A unique attribute of twigs is that they have bud scars, or **bud scale scars**, that mark where the terminal **bud scales** were attached before they fell away during the spring flush. These scars encircle the twig and appear as one or several rings around it. Because each scar represents where the terminal bud was at the end of a growing season, you can move along the twig and find older scars from years past. In this way, you can determine the age of a twig and judge how vigorous it was over the past few years (longer shoots usually indicate more vigor).

Twigs, as well as leaves, buds, and fruit, can be covered in tiny hairs (called **pubescence**) or a waxy substance called **bloom**. Think of the fuzz on a peach or the whitish coating on a grape, respectively. When pubescence is notably dense, it's described as **tomentose**. The twigs of many species feature pubescence when they first emerge in the spring, but only certain ones retain it through the summer. A twig covered in bloom is referred to as **glaucous**, and it generally has a whitish or bluish appear-ance. Pubescence and bloom are often easy to rub off, so in the later summer or winter, be cognizant that it may have simply been lost or reduced; check for other traits or come back in spring. If a twig (or a bud or a leaf) is naturally hair-less, it is **glabrous**.

Lenticels are gas-exchange pores in the bark that connect the inner tissue with the outside air. They appear on young twigs as small dots. As most twigs age, they begin to lose their lenticels, which all but dis-appear on older branches. Some species, however, like cherries and birches, retain their lenticels on thin bark as they age (a great identification characteristic, of course). Some lenticels are larger and more obvious than others—sometimes prominent to the point of looking warty— while other trees have very few or lack them altogether.

When **deciduous** trees lose their leaves in the fall, the twigs are left with marks where each leaf was attached, called **leaf scars**. Leaf scars are always directly below the bud, and they have consistent shapes and attributes dictated by how the leaf was attached. Some scars also have what look like dots within them; these are the ends of the **vascular bundles** that piped sugars and water to and from the leaf. The size and shape of a leaf scar, and the pres-ence, number, and pattern of the vascular bundles, can be excellent identification clues.

In the very center of a twig is the **pith**. It's not visible unless you cut the twig and look at the cross section or cut it through the center lengthwise. The pith is made of spongy tissue that can vary in color between species. Most trees have a solid pith, but a few have hollow piths or, in the case of the walnuts, chambered piths. As a diagnostic feature of a tree, checking the pith should be used sparingly, as it requires damaging a twig.

The most prominent, and often most important, components of the twig are the buds. **Buds** are bundles of embryonic tissue at the end of a growing tip and along the twig that will become new leaves, shoots, flowers, or a combination of these parts in the springtime. Think of them as next year's growth in miniature, put into storage in midsummer until the following spring.

All our trees have buds, and each species has buds that are unique to it. In fact, if you were so inclined, you could study only the buds and twigs of the trees in our region and still be able to identify almost all of them in the field down to the species without any other clues. In winter, a bud is

sometimes your only clue aside from the form, texture, and bark.

Two different types of buds are useful for tree identification: terminal buds and lateral buds. **Terminal** buds are set at the end of a twig, like a cap. **Lateral**, or **axillary**, buds are set along the length of the twig; these are the most important to take note of because they dictate the next step in the journey, as they define the leaf.

The paragraph you just read is probably the most important one to remember. If you take anything away from this section, remember that *a leaf is defined by its bud.* Once you understand this aspect of botany as it relates to trees, everything else is more or less derivative. You'll then be able to figure out the leaf arrangement and whether it's compound or not, which are two of the most important steps in the identification process.

Everything connected to the stalk that is just below the axillary bud is one whole leaf. The two essentially define each other: the axillary bud is the bud in the **axil** of a leaf, and a leaf is the organ that grows at the base of an axillary bud. (The term *axillary* simply refers to the position

BACKUP BUDS

A bud is always at the base of a leaf, you say? But what about dormant buds below the bark? Excellent question. You may have heard a gardener or arborist talk about **dormant** buds beneath the bark that begin to grow when a tree is wounded, creating **epicormic sprouts**, commonly called **water sprouts** or **suckers** (though the latter usually refers to sprouts that come from the base of the stem or roots).

These buds are more precisely called **latent** buds, essentially backup measures that trees produce in case their normal growing tips are damaged or otherwise removed. Their growth is triggered by certain hormonal cues from the rest of the tree (usually in response to damage like breakage or pruning), and they are physiologically distinct from the **vegetative** and **floral** buds that are easy to see along the twig. Latent buds are not readily visible and consist of undifferentiated stem cells rather than embryonic leaves or flowers. They are called "dormant" because they can lie dormant for many years—perhaps indefinitely—if they are not triggered to grow.

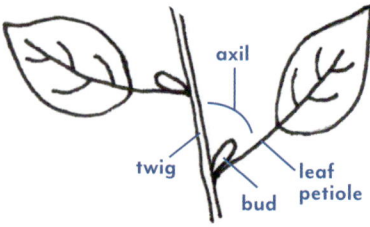

The buds in the leaf axil, the area between the petiole and the twig it's attached to, are known as axillary, or lateral, buds.

of a lateral bud being in the *axil* of the leaf, which is the angle between the leaf stalk and the twig that it is growing from.)

You may also hear a couple of other bud terms. A **pseudo-terminal** bud is a lateral bud that appears to be for all intents and purposes the terminal bud, when in fact the terminal bud has died or never grew to begin with. Some trees do this as a matter of course, like beeches (p. 330), sycamores (p. 352), and the honeylocust (p. 371). An **accessory** bud appears directly next to or below another bud, usually the terminal bud. They are very common and easy to see on ash trees (p. 188).

Lateral buds—referred to as such from here on out because sometimes the leaves will be absent, in which case the axil will not necessarily be obvious but can be assumed—are set in a consistent pattern along the twig called the **bud arrangement**. The spot where a bud is set is called a **node**, and the number of buds at a given node defines the bud arrangement. If there is just one bud per node, either alternating on either side of the twig or spiraling around it, then the bud arrangement is **alternate**. If there are two buds per node, the bud arrangement is **opposite**. An arrangement of three or more buds per node is called **whorled**.

As the leaf is what arises from just below the bud, the bud arrangement defines the leaf **arrangement**. They are equivalent terms: if the buds are oppositely arranged, the leaves will be oppositely arranged, and so will the entire branching pattern of the tree. Leaf (or bud) arrangement is the main clue to take away from twigs because you can split nearly all of our common trees into two categories: alternate or opposite. Sparingly few trees covered in this guide have whorled buds or buds that are **sub-opposite** (slightly offset sometimes, but mostly opposite). Thus, those attributes are defining ID characteristics.

Like twigs, the buds themselves have a range of attributes. One of the most prominent is their coverings, called **bud scales**, which surround the delicate embryonic tissues of next year's growth. Some buds have several scales that overlap one another, like shingles on a roof; these are called **imbricate** scales. Other buds have just two scales that open opposite each

LEFT: *Opposite arrangement* **RIGHT:** *Alternate arrangement*

Bud scale attributes from left to right: imbricate, valvate, naked, and single

other; these are called **valvate** scales. A few species are **single scaled**, with just one scale that unfurls like a sheath, or they have no scales at all, leaving their embryonic leaves to fend for themselves through the winter; the latter are called **naked** buds. Most trees in our area have imbricate scales, and the number of scales per bud can be quite variable. If you find a tree with distinctly non-imbricate scales, that is a good clue to its identity.

Other bud attributes to watch for include their angle of attachment; their size and shape; if they are appressed to the twig or not; if they're curved toward the twig; if they are blunt tipped, rounded, or pointed; their color, texture, and patterning; if they have a tiny stem or not; and if they have pubescence or any waxy coating. The same descriptive terms for twigs apply to buds.

When looking at the bud arrangement, make sure to look at a few different twigs and down the twig, not just at the very tip. Buds are set as new twigs are growing; the more vigorous the shoot, the farther apart the buds may be. So, on trees with low vigor for whatever reason, or near the end of a shoot where the growth naturally slows down, the buds may be tightly packed together, and they may *seem* to be opposite when they are in fact alternate. Also watch out for tight spirals of alternate buds; don't get them confused with opposite pairing. If they are opposite, it should be quite pronounced, such that you'll be able to see many examples.

Leaves

Leaves are of course the gold standard of tree identification. Many authorities consider them to be the most conspicuous component of a tree, so they instruct you to start with them when identifying a plant. In the Stepwise Journey, however, leaves are at best the third stop after

Bud types from left to right: conical, rounded, incurved, and stalked

A THORN, SPINE, OR PRICKLE?

Did you happen to see a thorn on that twig? Or was it a spine? Or maybe a prickle? Botanically, those are three distinct types of pokey growths because they are each modified from different parts of the tree. A **thorn** is modified stem tissue; it is woody, comes from a bud (or has a bud on it), can be branched or singular, and often appears as a short spur shoot. **Spines**, on the other hand, are modified leaves or stipules. They do not branch, and they arise from below a bud, just like a leaf. Finally, a **prickle** is a modified growth from the epidermis, or bark, of a plant. Think of it as a modified hair (called a **trichome**). Prickles do not have vascular tissue and are superficially attached (i.e., they do not go deeper than the outer bark).

taking in the whole tree, its surroundings, and its twigs. Now, you could argue that looking at the twigs is essentially the same as looking at the leaves, if there are in fact leaves on those twigs, especially if it's an evergreen tree or a conifer. But if it's a broadleaf tree, you'll need to find the buds on the twig before you can say with confidence what the leaf even is. Like I said, it's the third step, at best.

TOP: *Pinnately veined leaf*
BOTTOM: *Palmately veined leaf*

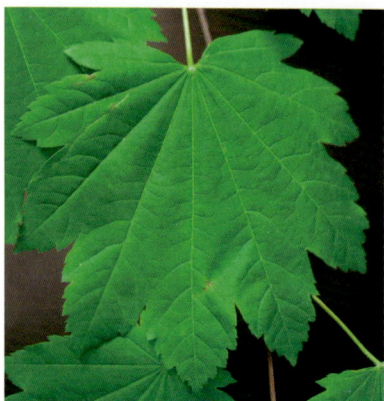

A **leaf** is the organ that arises from below the bud and generally performs photosynthesis. The second most important thing to remember after the bud (i.e. leaf) arrangement on broadleaf trees is that a whole leaf includes everything from the twig out. Conifer leaves are slightly different, so I'll focus on them after discussing the broadleaf trees.

The stalk of the leaf that attaches to the twig is called the **petiole**. At the other end of the petiole is the **blade** (also called the **lamina**), the surface of the leaf, which is interwoven with a network of veins. The extension of the petiole through the blade is called the **midvein**. The veins inside the leaf of a broadleaf tree split off, becoming smaller and smaller and creating patterns. Generally, veins split in two dominant patterns: **pinnate** or **palmate**.

Pinnately veined leaves maintain a central midvein that has side veins coming off to the left and right. Palmately veined leaves split into multiple different main veins where the petiole meets the leaf blade. You can remember which is which by thinking of the palm of your hand for *palmate*: your fingers all radiate out from your wrist like the veins radiating out from the petiole. *Pinnate* means "feather-like," so imagine the lateral leaf veins coming off the midvein like the barbs of a feather off the central shaft.

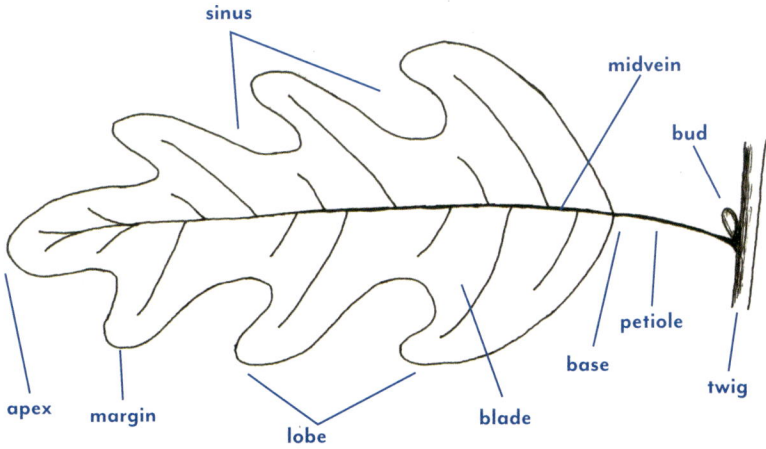

ABOVE: *Parts of a simple leaf* **BOTTOM LEFT:** *A trifoliate compound leaf*
BOTTOM RIGHT: *A pinnately compound leaf*

The vein pattern of a broadleaf tree often dictates the shape of the leaf. Pinnately veined leaves tend to be longer than they are wide or more heart shaped, whereas palmately veined leaves tend to be more rounded. The venation of a leaf is an important ID characteristic for two main reasons. First, certain species or genera have specific patterns and numbers of veins. How they are arranged, if they're curved or straight, if they're prominent or subtle, and how many there are can all be important clues. Second, if a

leaf is compound, the venation is what dictates what kind of compound leaf it is.

A **compound** leaf is one where the blade splits into two or more sections, each section called a **leaflet**. A leaflet does not have a bud at its base. If the leaf blade does not split into multiple, unconnected sections, then it's called a **simple** leaf. Compound leaves split into leaflets around their veins, such that if the veins are pinnately arranged, then the leaflets will be as well, making the leaf a pinnately compound leaf. The same is true

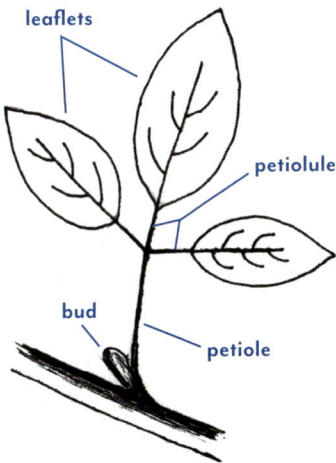

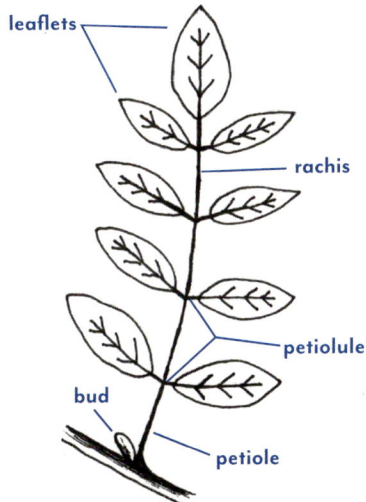

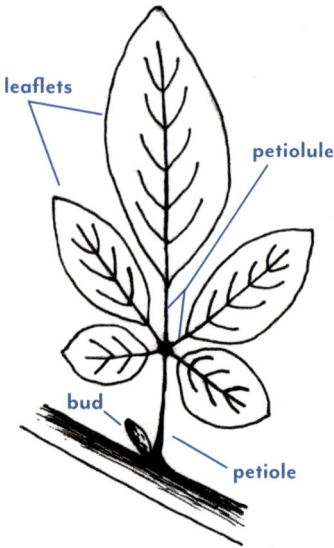

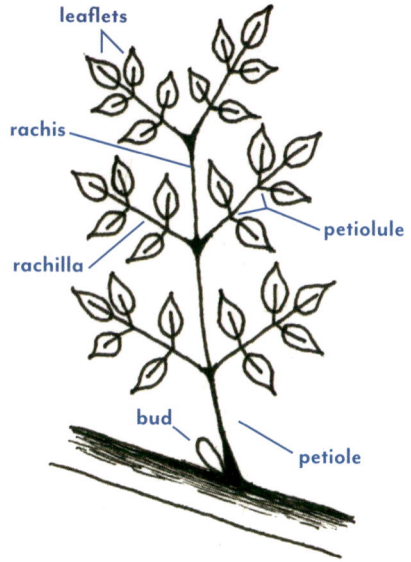

LEFT: *A palmately compound leaf* **RIGHT:** *A bipinnately compound leaf*

for palmately veined leaves: if they split into leaflets, then they are considered palmately compound leaves.

If the leaflets split again, then we add *bi-* to denote they split twice. We have a few species that have compound leaves that are **bipinnate** (e.g., honeylocust, p. 371, and silk tree, p. 372), but none that are bipalmate. If they split into only three distinct leaflets, they are called **trifoliate** (goldenchain tree, p. 374, is the best example).

When a leaf is compound, all the secondary parts get new names. The stalk of a leaflet is called the **petiolule**, which just means, adorably, a smaller petiole. If a leaf or leaflet does not have a petiole, it is called **sessile**. (When *ule* is added to a descriptive term, whether in the middle or at the end, it denotes it's a miniature version of whatever is being described; you'll notice it with the descriptions of leaf margins.) Each leaflet in a pinnately compound leaf is called a **pinna** (plural: pinnae). The central stalk that the pinnae are attached to is called a **rachis**; it is equivalent to the midvein on a simple leaf. If the leaf is bipinnately compound, the central stalk in each pinna coming off the main rachis is called a **rachilla**, to which the leaflets are attached.

Determining whether a leaf is compound or not is one of the primary reasons that finding the buds is so integral to tree identification. Mistaking a simple leaf for a compound leaf will throw you off course almost immediately. On the other hand, once you determine whether

The apex of a leaf is the tip opposite the petiole. Leaf apex types, left to right: acuminate (concave sides), acute (flat sides), emarginate (notched), rounded (rounded), truncate (flat)

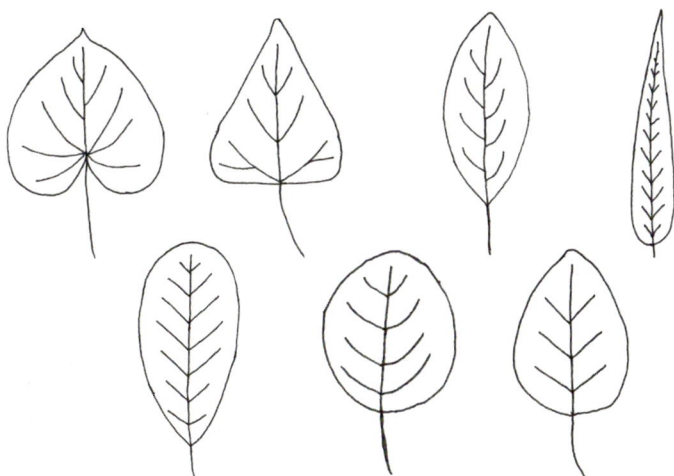

Overall leaf shapes, top row: cordate (heart shaped), deltoid (triangular), elliptic (oval), lanceolate (long and skinny); bottom row: obovate (widest near the tip), orbicular (circular), ovate (widest near the base)

the leaves are alternate or opposite, followed by whether they are compound or simple, you will have narrowed down your options substantially. The Stepwise Journey is fundamentally an ordered process of elimination, like the classic game Guess Who? If your tree has oppositely arranged leaves, then you can eliminate all the alternately arranged ones, and if those leaves are compound, then you can eliminate all the trees with oppositely arranged simple leaves. Now you have only a few species left to consider, and you can move on to focusing on the different attributes of the leaflets or other ID characteristics that separate them.

Leaves are described using their general shape, but they are also broken down by section. There are terms to describe the leaf tip (or **apex**), the leaf **base**, the

top or bottom surface, and the outer edge or **margin**. Any given tree may have any combination of attributes, and trees that are closely related tend to have similar leaf types. Be wary, though, as similar-looking leaves may be a superficial characteristic two unrelated trees have in common. It's the flowers and fruit that define who is related to whom, but more on that later.

Leaf shape, one of the first and most obvious traits to consider, can either be described as an overall shape or broken down into parts, like the tip and the base. Every shape has its own technical term, and many have their own modifiers as well. For example, the prefix *ob-* modifies a term to mean the inverse of its normal definition: while **ovate** means egg shaped, where the wider part of the leaf blade is

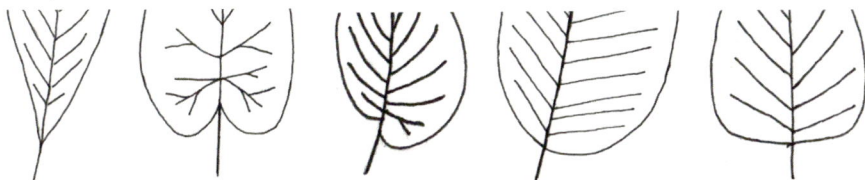

The base of a leaf is the end that has the petiole attached. Leaf base types, left to right: cuneate (wedge shaped), cordate (heart shaped), oblique (offset), rounded (rounded), truncate (flat)

The margin (outer edge) of the leaf. Margin types, top row: entire (smooth), crenate (scalloped), dentate (sharply pointed outward); bottom row, serrate (sharply pointed forward), doubly serrate (larger serrations that are themselves serrated). The addition of ule *plays a similar role here as it does for the term* petiole, *such that serrulate, denticulate, or crenulate refers to miniature versions of the original.*

near the base and the skinnier part is near the apex, an **obovate** leaf is just the opposite, where the wider part is near the apex and the skinnier part is near the base.

As with the surface of twigs and buds, the surface of a leaf can vary. Most of the same terms and definitions apply: a pubescent leaf is fuzzy, a glabrous leaf is hairless, and so on. Similarly, leaves can have a waxy bloom on them, especially on the underside, though it's not uncommon to have bloom on both sides, especially for conifers. There are many different technical terms to describe textures, such as **rugose**, **coriaceous**, and **scabrous**, but often the nontechnical terms are more helpful: wrinkled, leathery, and rough, respectively. Keep in mind that these textures can appear in combination and may require close inspection to confirm. For example, a leaf can be rough but hairless, or smooth yet covered with hairs. Finally, some leaves are evergreen and tend to exhibit traits that differ from deciduous leaves—often they are thicker, leathery in texture, and have a tough, waxy outer coating.

Conifer leaves have their own attributes that are slightly different from those of the broadleaf trees. They often seem more confusing because they don't always conform to such simple rules and divisions as broadleaf trees do, and their shapes are very different. But as with broadleaf trees, conifers' characteristics can be broken down to help you separate them out and narrow down species. Start with the leaf type; the conifers in our area have leaves of three main types: needle-like, scale-like, and awl-like.

Needle-like leaves (often just called needles) are the most common conifer leaf type that you'll find in the Pacific Northwest. Needles are skinny, stand away from the twigs, and are much longer than they are wide. Needles of different species, however, vary greatly in how they are attached to the twig and how they spread from it.

Some needles are **singly borne**, where just one needle arises from a point along the twig. These attachments are most often spirally arranged along the twig.

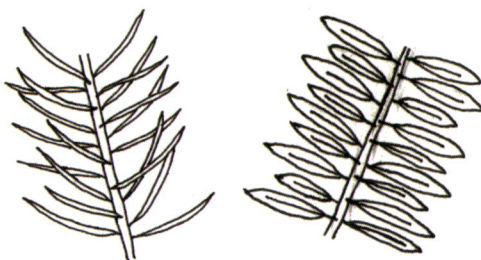

LEFT: *Single, rounded conifer leaves* **RIGHT:** *Single, flat conifer leaves*

TOP: *A decussate orientation produces a plus shape when viewed from the tip.*
BOTTOM: *The needle-like leaves of Pacific yew are two-ranked.*

Needles are also borne in bundles or groups. In the Northwest, all the species of pine in the genus *Pinus* have needles in bundles of either two, three, or five, connected at the base, so they are obviously bundled together in discrete groups. These are called **fascicles**, and if you see these, then you know for sure you've found a pine; no other tree has needles in fascicles of two, three, or five. Sometimes many more than five needles grow from what appears as a distinct peg. These needles are technically singly borne but on very slow-growing **spur shoots**. Cedars (*Cedrus*) and larches (*Larix*) have this needle arrangement.

Now, even though they are called needles, most are not perfectly round or sharp. Needle leaves for most species are oriented with distinct upper and lower surfaces, but their unique characteristics don't stop there. When viewed in cross section, individual needles may be flattened, rounded, square, or triangular. If it's a pine, they'll be split into sections of a circle: pine needles begin as essentially one round needle that splits into halves, thirds, or fifths, like slices of pie. In addition, needles can have grooves or ridges on any of their surfaces, they can twist around, and their tips can be pointy, flat, or notched.

An important attribute of needles (and an easy way to tell the bottom from the top for some species) is stomatal bloom. **Stomata** are tiny pores in the leaves that are used in the exchange of gas and water.

However, the needles themselves may not follow that pattern. Some needles grow outward in all directions like a bottlebrush, while others grow in distinct **ranks**, where all the needles, regardless of where they are attached to the twig, grow along distinct planes. Several trees in our region have **two-ranked** leaves, meaning their needles grow along two planes, usually one to the left of the twig and one to the right. Twigs with leaves growing in three distinct planes are **ternate**, while those growing in four are called **decussate**.

Needles groups, left to right: fascicles of two, three, and five, and a spur shoot with many more

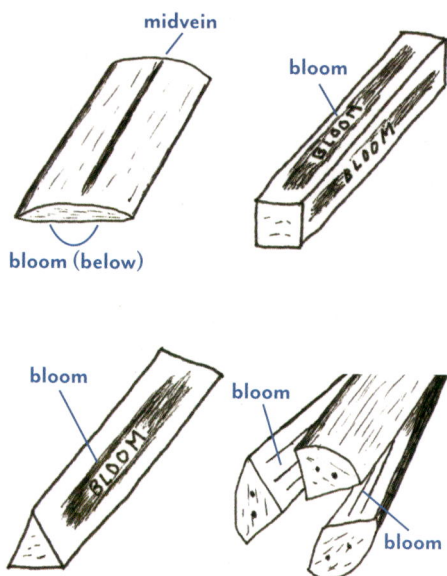

Cross section of conifer leaves, clockwise from top left: flat, square, pie, and triangle

Stomatal bloom is a waxy substance produced by the leaf that appears as white markings or lines, most often on the lower side of the leaf, but not always. If you look closely enough, the pattern of the bloom itself can be helpful too. For example, some leaves have two distinct lines on their undersides, while others have several distinct lines or none at all.

Finally, some needle-bearing trees in this region are **deciduous**, their leaves falling away each autumn. These leaves lack the protective waxy coatings that **evergreen** leaves have, their texture is far softer to the touch, and they are generally a lighter shade of green compared to their evergreen counterparts. It's worth noting that *evergreen* simply means that a tree keeps its leaves for at least one full year, though often they'll keep them over multiple seasons. In time, evergreen trees shed their oldest leaves—those that are farthest down the twig. So, for example, if a pine tree keeps its needles for three years, at the end of its fourth growing season it will shed its needles from the first

year, keeping only the three most recent years' worth of foliage.

Scale-like leaves (often just called **scales**) are the second most common conifer leaf type in our region, and if you find them in the field, you can be certain they belong to a tree in the cypress family. Scale-like leaves are very small and **appressed** to the twig, as if fused with it, so it appears that the leaves and the twigs are one and the same. However, when you look closely, patterns emerge as a result of their arrangement, shape, and whether they overlap or not; start here when trying to narrow down a species with scale-like leaves.

Scale-like leaves often grow in opposite pairs or in threes. When they are in pairs, they tend to have **dimorphic** leaf growth. *Dimorphic* means "two forms" and refers to the same part growing in two different, distinct ways. In the case of scale leaves, one pair grows on either side of the twig, and another pair grows on the top and bottom, each pair having a certain shape based on its placement. This is repeated down the twig, creating a recognizable pattern unique to each species, often with leaves slightly overlapping each other. Their pattern stands out most clearly when you turn the twig over and look at the underside. Like needles, scales can have stomatal bloom, and it contrasts with the green of the leaves such that, when considered along with the shape and size of the leaves, it is a very good identification characteristic.

The third common leaf type is **awl-like**. An awl is a hand tool like a small pick, where the blade starts wide and narrows down to a point. For conifers, this means the base of the leaf is wide and attached to the twig in the same manner as a scale leaf, but instead of being appressed to the

From left to right: Scale-like (does not overlap), scale-like (does overlap), awl-like

twig, the rest of the leaf bends outward and becomes pointed and more like a needle. Awl-like leaves can be seen as an in-between of scale-like and needle-like, and indeed scale-like leaves often have reflexed tips, which illustrates how these descriptions fall along a gradient.

If you find a tree with awl-like leaves, it's almost certainly one of a few species in the cypress family, and the same arrangement patterns you can see for scale-like leaves can be seen with awl-like leaves. Some species (specifically the junipers) can have both leaf types on the same twig, while others that normally have scales may have cultivars that have only awl-like leaves.

Flowers, Fruit, and Cones

The reproductive parts of trees (indeed of all plants) are used as the basis of the scientific taxonomic system. It turns out that the reproductive parts are very consistent within a species, and by tracing their similarities and modifications, taxonomists can find evolutionary relationships. As these traits are fundamental to how species are differentiated, some familiarity with flowers, fruit, and their parts is important for tree identification.

Flowers are the reproductive organs of broadleaf trees (except for the ginkgo, but more on that

later). Flowers are unique organs on trees, and as such have their own terms. The stalk that holds a flower is called a **peduncle**, and the base of the flower that it attaches to is called the **receptacle** (just as with leaves, if the flower lacks a peduncle, it is sessile). The receptacle holds the rest of the flower, which usually has four main parts: sepals, petals, pistils, and stamens. **Sepals**, the outermost appendages that cover the emerging flower, can be found at its base when it opens. The collection of sepals is called the **calyx**. **Petals** are the next layer of appendages; they're usually the showy parts that draw our attention (and the attention of pollinators, of course). Collectively they

Parts of a flower

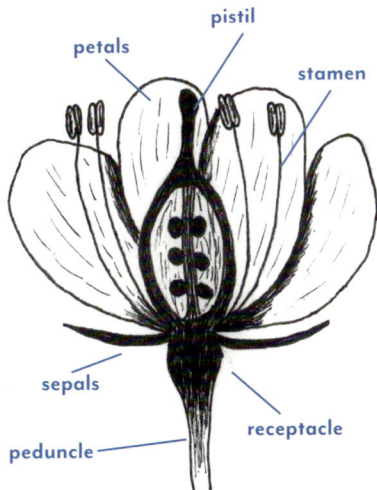

43

are called the **corolla**. (When there is no clear distinction between sepals and petals, they are all called **tepals**. This is the case for magnolias.) **Pistils** are the female parts that hold the **ovules** (embryonic, unfertilized reproductive cells; think of an egg cell) and receive **pollen**. From the top down they're composed of the **stigma**, **style**, and **ovary**. The male parts are the **stamens**, and they produce the pollen (male reproductive cells). On the top of each stamen is the **anther**, which is held up on the **filament**. Finally, some trees also have **bracts**, which are modified leaves that appear and act as petals, such as on some dogwoods.

All flowers are simply different combinations, modifications, and arrangements of these parts. A flower that has stamens, pistils, petals, and sepals is a **complete flower**. Flowers that have at least both female and male parts in the same flower (whether they have petals and sepals or not) are **bisexual flowers** or **perfect flowers**. **Unisexual flowers** are either female (**pistillate**) or male (**staminate**) and only have the corresponding parts. These can be produced separately on the same tree or on different trees. If a tree produces both types of unisexual flowers on the same tree, it is **monoecious**, which means "one house." If a species produces only pistillate or staminate flowers on separate trees, it is **dioecious**, meaning "two houses"; in this case, trees are called either female or male.

Flowers can be attached singly or in a group, called an **inflorescence**. An inflorescence is essentially a compound flower structure that includes all the flowers in a discrete grouping along with their stalks, called **pedicels** (akin to the petiolules in compound leaves). The arrangement of the individual flowers in an inflorescence can help determine a species, and for the purposes of this guide, the inflorescence is more important than the anatomy of the flowers themselves.

Flower arrangement also dictates the arrangement of the fruit. Botanically, a **fruit** is a mature ovary, so however the flowers with the ovaries are arranged, the fruit will necessarily follow that pattern; those in groups are called an **infructescence**. All flowering species produce some kind of fruit, but it's not always the type of fruit people think of in a culinary sense.

Fruit takes many different forms, so it's one of the best ways to nail

Inflorescence types, top row, left to right: spike, raceme, panicle; bottom row, left to right: corymb, umbel, cyme

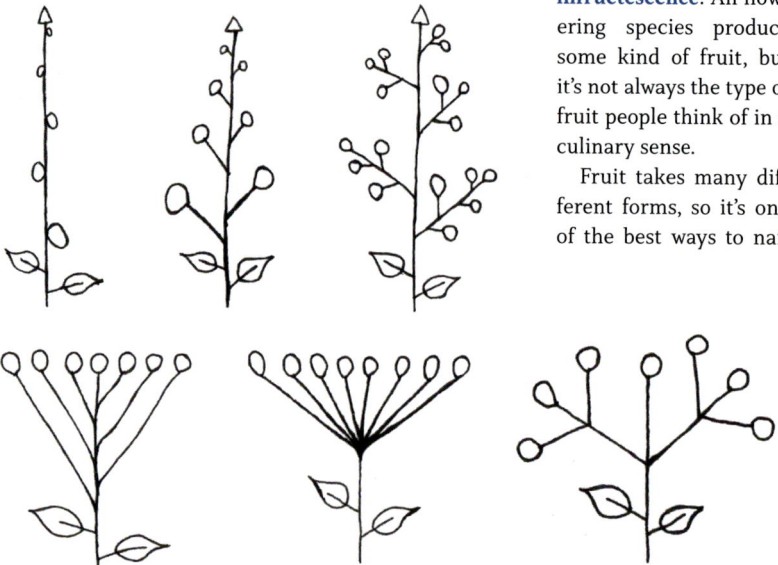

Some fruit forms, from left to right: the dry fruit of hop-hornbeam (a loose, papery catkin),
fleshy fruits of Lavalle hawthorn (a collection of small pomes), dry capsule fruit of Pacific willow
(capsules that split open to release fuzz-covered seeds)

down a species. Unlike flowers, though, fruit is not necessarily indicative of an evolutionary relationship. That is, not all berries or nuts come from closely related trees that passed down that fruit type. Fruits evolve based on circumstances like environmental and dispersal pressures, and a tree that has a fleshy fruit today may have evolved from a lineage with dry fruit, or vice versa. However, like flowers, fruit is very consistent across individuals of a given species and thus is a reliable identification characteristic.

Botanically, fruit can be challenging— but it's a fruitful rabbit hole to go down; the tiniest anatomical details can make a difference. For example, the fleshy parts of several succulent fruits may look superficially alike, but they may have developed from different parts of the flower structure, like the receptacle, ovary, or seed coat. The same can be said for dry fruits. For the purposes of this guide, paying attention to general fruit characteristics is enough to identify a species when you consider it alongside other aspects.

A genus commonly has a specific fruit type that all its species share, with slight variations between them. For example, maples all have double samaras, but each species has its own unique variation on this design (see table on p. 169). The same

can be true for families as well, like the pea family, Fabaceae, whose fruits look similar across the board. However, there are always exceptions: the fruit of the rose family, Rosaceae, varies greatly across genera and species.

Cones are the reproductive parts of conifers; they differ from flowers in that they do not have ovaries or other specialized sexual parts. Cones come in two

Pollen cones of an Atlas cedar

LEFT: *Seed cones of a Jeffrey pine* MIDDLE: *A classic cypress family cone (an Arizona cypress cone)* RIGHT: *A classic pine family cone (a Torrey pine cone)*

types: pollen cones and seed cones. **Pollen cones**, as their name suggests, produce pollen; they mature most often in winter and early spring. Pollen cones are small, flimsy, and hardly noticeable in most species until they release their pollen. Once they release their plumes to the wind, they fall away, disintegrating quickly. (Seed cones, on the other hand, are what everyone generically calls "pine cones," though this is often far from accurate.)

Seed cones can be superficially understood as the fruit of conifers, though of course they are botanically quite different from the fruit of broadleaf trees. Like broadleaf flowers and fruit, cones are the primary means of separating out species and genera of conifers. In contrast, however, each species of conifer has its very own unique cone. This means that a pine cone comes only from a pine, a spruce cone from a spruce, a larch cone from a larch, and so on. In fact, if you practice enough, you could identify nearly all our common conifers by the cone alone.

The two most common conifer families in the Pacific Northwest are the pine family (Pinaceae) and the cypress family (Cupressaceae). These two families can be easily separated by the traits of their cones. Trees in the pine family have cones that are elongated, with a central axis covered in spirally arranged, overlapping **scales**. Though technically all the pine cones' scales have associated **bracts**, they are not always present or obvious. Most cypress family cones, on the other hand, are more spherical and have **peltate** (shield-like) cone scales that abut each other.

When it comes to anatomy, a few oddball species throw convention out the window, creating confusion out in the field. The yews, for example, have all but lost their cones, having reduced them to just a single seed surrounded by a red **aril**, which looks like a red berry; likewise, the junipers' cones have been modified over time to appear as succulent, blue berries. Don't even get me started on the umbrella-pine—it isn't even in the pine family but looks like them in almost every way. Luckily, these strange trees have plenty of attributes that help them stand out when you move through the Stepwise Journey.

Bark

As you move along the Stepwise Journey, you look at all aspects of the tree and, based on what you see, eliminate options that don't fit. As you slowly eliminate possibilities, you'll likely start to lean toward a certain species, but you'll need a few

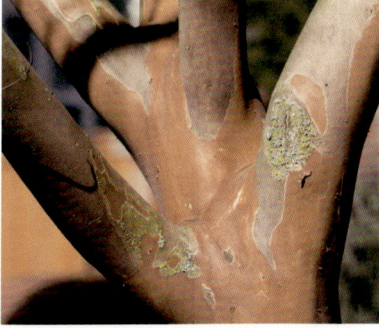

Bark types, clockwise from top left: smooth; rough, but not furrowed; irregularly ridged; regularly ridged; covered in large bark plates; exfoliating like puzzle pieces; exfoliating in strips; fibrous

THE EXCEPTIONAL GINKGO

You can forget everything that was just explained about broadleaf trees and conifers when it comes to the ginkgo. Ginkgoes are of an ancient lineage that extends back over 290 million years, predating any broadleaf tree alive today and most of the conifers too. They have broad leaves, but they do not have ovaries or seed cones, so they cannot be classified with either conifers or broadleaf trees—they truly are in a class of their own (figuratively and botanically speaking).

So if you run across a tree that doesn't seem to follow any of the patterns described here, you're probably looking at a ginkgo. Luckily, they are so well-known and unique that they are hard to confuse (see p. 348).

more details to confirm your hunch. Bark is often either the attribute that tips me off toward a species or the part that finally confirms my hunch.

Bark's usefulness for tree identification is really on a spectrum. Some bark is so striking and unique that it is all you need to identify the tree. Other times, bark can be so generic that it's all but useless. On top of that, bark can change as a tree gets older, starting out smooth and mundane, then transitioning into a wildly intense pattern with age. Luckily, no matter the species, bark is usually consistent across individuals and thus can be used as a helpful added attribute.

Some species have bark that is distinctly theirs, ginkgo being a good example. Other species share common bark traits with their closest kin and so are set apart, like spruces. When looking at bark, consider the color, the texture, and traits like curious shapes, lines, or thorns.

Just as with trees' textures and overall forms, your mind can discern the slightest nuances in bark patterns; as with different types of wallpaper, your eyes can easily pick out unique patterns, but noticing what exactly is different about them may be more of a challenge. It's useful to describe the bark to define what makes the pattern stand out. Look intentionally and take the time to describe specific

attributes to apply definitions to what you're seeing.

Putting It All Together

Now that you have learned all the technical aspects of trees and tree parts, the real fun begins. You can start to apply your knowledge to the trees around you, essentially putting faces to the names. I recommend that you simply head outside and start looking at the trees around you. Snag a twig and look at it. See if you can start to pick out the different parts of a leaf or a flower. Taking a close look at a few trees will help you begin to see the differences between them. Just observe; you don't necessarily need to identify them, but why not? This is a perfect opportunity to practice the Stepwise Journey.

First, step back and take in the whole tree. Observe its texture, color, shape, size, and architecture. There isn't any wrong way to go about this, but try to make your process repeatable. Take a look at nearby trees to see their differences. Does one have very upright limbs with a central leader, while another grows more pendulously? Are they more excurrent or decurrent? Does it have a rough, harsh texture, or is it feathery and wispy? Any aspect can be useful.

Once you're under the tree with a twig in hand, look for the buds to determine first if they are alternately or oppositely arranged; then, find where a leaf starts so that you can determine if it's simple or compound. If there are no leaves, examine the twigs and the buds themselves for clues: Is anything particularly striking about them, like a color or pattern or shape? Are the twigs stout, suggesting perhaps the leaves are large? Are there old parts still hanging on, like fruit or the rachises of old compound leaves? Are the twigs short, straight, curvy, or long?

It can be daunting to start with a twig and no leaves, as in winter, so don't despair if figuring out the species takes a little while. Even experts have trouble with winter identification. Flip through the guide to see if you can find a few twigs that look similar, then compare their specific traits to your tree. You can focus on the small details, but also trust your eyes. When you think you have a match, or want to eliminate a close look-alike, it's time to move to the next step: examine the bark, the form, and any fruit or flowers, as well as the context of where your tree is growing.

If there are leaves on the tree, your job is a little easier. Once you determine whether they are simple or compound, your next step is to look closely at the leaf's characteristics, such as its margin, thickness, hairiness, size, and overall shape. Using this guide, flip to the proper section based on the leaf type and arrangement, then peruse the entries for leaves that match. Again, trust your abil-ity to recognize a potential match. If you find one that looks similar to your tree, read the description and see if the details line up. Eliminate look-alikes or confirm your species by examining other traits: when you pin down the right tree, everything should line up just right.

If a tree profile seems to match the tree in front of you, odds are you've identified that tree—well done! You've also looked more closely at a tree than most people ever will in their lives. In doing so, you've built a mental profile of that species; you'll probably start to notice it more often around you, just like when you learn a new word and then begin to hear it all the time. As you see a tree more often, you'll start to recognize it out of hand, and you'll get faster until, just like that, you're like a magician, identifying trees by their outer contour from blocks away.

I bet you'll also start to notice the trees that you aren't familiar with, ones that don't fit the patterns you've already learned to recognize. The Stepwise Journey is designed to help you focus your attention on different details and notice when they diverge from what you expect, in an organized, repeatable way. Ideally, you'll be able to return to this guide, find the species you thought it was, and learn that there's a closely related species with just that trait you found—another successful ID. If not, then you'll be armed with all the technical knowledge to navigate another, farther-reaching resource with confidence, building your skills and local knowledge every step of the way.

Sometimes a Unique Trait Is All You Need

Once you are familiar with the trees of your area, you may begin to find that a single attribute can give a species away, some kind of signature trait like cobalt blue fruit or peeling bark. But if you aren't familiar yet with which trees have which traits, it can be frustratingly tantalizing to see some unique trait that clearly stands out yet have no decent way to connect it to a certain species. (This is when I used to search for terms like "tree with peeling bark, PNW" on the internet and immediately get overwhelmed with options.) These lists are my attempt to address this issue. If some trait jumps out at you and you want to look only at the trees that have that particular trait, check whether I've included it here.

Refer also to the calendar of flowering times, which indicates when a tree with showy flowers may be blooming, for how long, and what color its flowers are.

GENERAL FRUIT TYPES AND COLORS

Berrylike: Yellow, Red, Orange

Apples and crabapples
Apricot
Autumn-olive (also green)
Cherry: Birchbark, bitter, Higan (often fruitless), Japanese flowering (often fruitless), sweet cherry (also black), Yoshino (often fruitless)
Chinese pistache (also blue/black)

Dogwood: cornelian-cherry, eastern flowering, kousa, Pacific
English holly
Ginkgo
Hackberry
Hawthorn: Autumn Glory, cockspur, midland, Lavalle, singleseed, Washington, Winter King green
Mountain-ash: American, European, Korean

Mulberries (also white)
Pacific madrone
Peach
Pears: common edible variety
Persimmons
Photinias
Plum: common varieties
Quince (large fruit)
Strawberry tree
Whitebeam
Yews

Berrylike: Blue, Purple, Black

Black hawthorn
Black tupelo
Cascara
Cherry: European bird, Sargent, sweet (also red)
Chokecherry: Amur, common

Common fig (also green)
Corktrees
Crabapples (some)
Fringetrees
Giant dogwood
Harlequin glorybower (with pink sepals)

Laurel: English, Portuguese, sweet bay
Olive
Plum: cherry, common (also yellow)
Sassafras
Serviceberries
Windmill palm

Berrylike: Brown, Beige, Green

Camellias (rarely fruit)
Common fig (also purple)
Dove tree

Pear: Asian, Callery, Ussurian
Oregon-myrtle

Russian-olive (also yellow)
Snowbells

Dry, Hanging Catkin

Alders
Aspens
Birches

Cottonwoods
Hop-hornbeam
Hornbeams

Poplars
Wingnut
Willows

Dry Capsule (Papery or Hard)

Carolina silverbell
Crepemyrtle
Eucalyptuses (gums)
Golden raintree
Japanese tree lilac
Katsura

Lindens
Paulownia
Persian parrotia
Rhododendrons
Seven sons flower
Snowbells

Sourwood
Stewartias
Willows
Witchhazels
Zelkova

Long, Hanging Pod (Round or Flat)

Black locust
Catalpas
Goldenchain tree
Honeylocust

Japanese pagodatree
Judas tree
Kentucky coffeetree
Maackias

Redbud
Silk tree
Yellowwood

Spiky Ball

Beeches (often split
  open)
Chestnuts

Golden-chinquapin
Horsechestnuts and
  buckeyes

Planetrees and
  sycamores
Sweetgum

Papery Samaras

Ashes
Elms

Hardy rubber tree
Maples

Tree of heaven
Tuliptree

Hard Nut (Sometimes in Husk)

Almond (in green
  husks)
Beeches (in spiky husk)
Butternut (in smooth
  green husks)

Filberts and hazelnuts
  (in brown husks)
Hickories (in smooth
  green husks)
Horsechestnuts and
  buckeyes (in beige
  husks)

Oaks
Tanoak
Walnuts (in smooth
  green husks)

Lumpy Mass with Seeds

Magnolias
Smoketrees (airy mass)

Sumac

Tuliptree (made of
  samaras)

UNIQUE BARK CHARACTERISTICS

Mostly Smooth, White

Cottonwoods (when young)
European white birch
Himalayan white birch
Paper birch and kin
Quaking aspen
White poplar

Mostly Smooth, Gray

Beech
Common fig
Crabapples
Douglas-fir (when young)
European hornbeam
English holly
Eucalyptuses (some)
European aspen
European mountain-ash
Flowering ash
Japanese maple
Magnolias (most)
Oak: bamboo-leaf, pin, silverleaf
Pacific dogwood
Red alder
Silk tree
Snowbells
True firs (when young)
Yellowwood

Exfoliating

Amur chokecherry
Birchbark cherry
Birches
Black hawthorn
Chinese persimmon
Dove tree
English hawthorn
Fringetrees
Hickory: shagbark, shellbark
Hop-hornbeam
Japanese tree lilac
Maackias
Pacific madrone
Paperbark maple
Seven sons flower
Silver maple
Strawberry tree

Camouflage Pattern

Crepemyrtle
Cypress: Arizona, modoc
Kousa dogwood
Lacebark elm
London planetree
Persian parrotia
Stewartia: Japanese (Korean), tall
Trident maple
Winter King green hawthorn
Yews
Zelkova

Corky or Warty-Looking

Common hackberry
Cork oak
Corktrees
English walnut

UNIQUE TWIGS

Petiole Covers over Buds

Bigleaf snowbell
Corktrees
Honeylocust

Japanese pagodatree
Planetrees and
  sycamores

Yellowwood

Corky Growths

Blue ash
Bur oak

Crepemyrtle
English elm

Hedge maple
Sweetgum

UNIQUE LEAF CHARACTERISTICS

Common with Purple or Reddish Leaf Varieties

Catalpas
Cherry plum
Common chokecherry
Crabapples
European beech

Japanese flowering
  cherry
Katsura
Maple: Japanese,
  Norway, sycamore

Redbud
Smoketrees
Silk tree

Common with Variegated Leaf Varieties

Boxelder
Cypresses (some)
Deodar cedar

English holly
European beech
Kousa dogwood

Junipers (some)
Willows (some)

Common with Cutleaf Varieties

European alder
European beech

European white birch
Japanese maple

Staghorn sumac

CALENDAR OF FLOWERING TIMES

This phenology calendar shows when different trees bloom across the seasons. If you see a tree with showy flowers at a certain time of year, refer to the column titles to find the season that fits best, then look down that column to see which trees are blooming at that time. The color of the bar corresponds to the color of the blooming flower. Once you find a few potential species, head over to the profiles to see if they match—ID made easy!

SPECIES	LATE WINTER	EARLY SPRING	MID SPRING
Camellia japonica	▬		
Hamamelis x intermedia	▬		
Parrotia persica	▬		
Cornus mas		▬	
Acer rubrum		▬	
Acer saccharinum		▬	
Prunus × blireana		▬	
Prunus cerasifera 'Atropurpurea'		▬	
Prunus cerasifera 'Thundercloud'		▬	
Prunus x subhirtella		▬	
Prunus x yedoensis 'Akebono'		▬	
Prunus x yedoensis 'Yoshino'		▬	
Magnolia stellata		▬	
Magnolia hybrids		▬	
Magnolia x soulangeana		▬	
Pyrus calleryana			▬
Prunus amygdalus			▬
Prunus domestica			▬
Prunus persica			▬
Prunus avium			▬
Pyrus communis			▬
Prunus sargentii			▬

CALENDAR OF FLOWERING TIMES

SPECIES	MID SPRING	LATE SPRING	EARLY SUMMER
Amelanchier arborea	▭		
Amelanchier x grandiflora	▬		
Amelanchier laevis	▭		
Cercis spp.	▬▬▬		
Prunus serrulata	▬▬		
Prunus virginiana	▬▬▬		
Malus domestica	▭		
Malus spp.	▬▬▬		
Malus spp.	▭▭		
Prunus laurocerasus	▭▭		
Prunus padus	▭▭		
Arbutus menziesii	▭		
Cornus florida		▬▬▬	
Cornus nuttallii		▭	
Crataegus monogyna		▭	
Davidia involucrata		▭	
Halesia carolina		▬	
Magnolia acuminata		▭▭	
Malus tschonoskii		▬	
Amelanchier alnifolia		▭	
Crataegus douglasii		▭	
Crataegus laevigata		▬▬	
Aesculus hippocastanum		▭	
Aesculus x carnea		▬	
Aesculus pavia		▬	
Acer tataricum		▬	
Acer buergerianum		▬	

SPECIES	MID SPRING	LATE SPRING	EARLY SUMMER
Fraxinus ornus		▢	
Laburnum × watereri		▬	
Paulownia tomentosa		▬	
Chionanthus retusus		▢	
Cornus controversa		▢	
Styrax obassia		▢	
Cornus hybrids			▬
Cotinus coggygria			▬
Crataegus crus-galli		▢	
Crataegus x lavalleei			▢
Liriodendron tulipifera			▬
Chionanthus virginicus			▢
Cladrastis kentukea			▢
Prunus lusitanica			▢
Robinia pseudoacacia			▢
Styrax japonicus			▢
Styrax japonicus 'Pink Chimes'			▬
Syringa reticulata			▢

SPECIES	EARLY SUMMER	MID SUMMER	LATE SUMMER
Cornus kousa	▢		
Stewartia rostrata	▢		
Catalpa speciosa	▢		
Crataegus phaenopyrum	▢		
Magnolia grandiflora	▢		
Stewartia monadelpha	▢		
Stewartia pseudocamellia	▢		
Tilia platyphyllos	▬		

CALENDAR OF FLOWERING TIMES

SPECIES	EARLY SUMMER	MID SUMMER	LATE SUMMER
Catalpa x erubescens	▬		
Tilia cordata	▬		
Catalpa bignonioides		▬	
Catalpa ovata		▬	
x Chitalpa tashkentensis 'Pink Dawn'		▬▬	
Maackia hupehensis		▬	
Tilia americana		▬	
Tilia tomentosa		▬	
Albizia julibrissin		▬▬	▬
Koelreuteria paniculata		▬	
Maackia amurensis		▬	
Oxydendrum arboreum		▬	

SPECIES	MID SUMMER	LATE SUMMER	EARLY FALL
Lagerstroemia spp.	▬▬	▬▬	▬▬
Lagerstroemia spp.	▬▬	▬▬	▬▬
Clerodendrum trichotomum		▬▬	▬▬
Heptacodium miconioides		▬▬	▬▬
Styphnolobium japonicum		▬▬	▬

CONIFERS

Conifers are the best, and I'm fine admitting that. It's been said that conifers all look the same, that it takes a botanist to tell them apart. Well, that notion ends today because you're going to learn all the secrets that Big Botany doesn't want you to know. It is my hope that you'll come to see how different they are and agree they should always come first too.

As you learned in the previous sections, conifers have just as many little parts and attributes that set them apart as broadleaf trees do. However, conifers in the Pacific Northwest, and certainly those in towns, yards, and parks, are in fact simpler to differentiate and identify than broadleaf trees if you know what to look for. There are several reasons for this: First, there are fewer of them to choose from, and they also lack the spectacular variation in parts like leaves and fruits that the broadleaf trees have. But the biggest reason is that they lack flowers altogether and simply have cones.

At the risk of oversimplifying, imagine cones as the flowers and fruit rolled into one. Botanically, this is a dubious statement, but functionally, it's a helpful framework for understanding how conifers are taxonomically organized. For broadleaf trees, taxonomists first use the flowers and then the fruit to split them up and group them together. Conifers don't flower and then produce fruit: it's a cone all the way down (though miniature at first).

Most cones look unique to their species, but if you happen to come across one of the few that closely resembles another, our conifers have graciously given us foliage and canopies that are—in all cases but a sparing few—evergreen and distinct. In combination, the form, foliage, and cone will reveal the exact species almost every time. About half the time, you will find that only one or two key attributes is enough with the Stepwise Journey to help you identify the species—

and I haven't even talked about the bark yet!

If you're using the Stepwise Journey for conifers, first, take in the whole tree and notice its form and texture. Second, look at the leaves and see if they are scale-like, awl-like, or needle-like; after some practice, you'll probably be able to tell this from the texture alone. Third, narrow down options based on the leaf arrangements and patterns you see.

Now, when you come across a conifer, you'll probably know innately: it'll look like what people often casually call a cedar or a pine or a fir. However, those generic, vernacular names are usually misleading when it comes to identifying a tree to its botanical species. For example, trees with scale-like leaves are often called cedars (think western redcedar, incense-cedar, and Port Orford-cedar). But botanically, they are not cedars at all. The same goes for Douglas-fir and umbrella-pine: the former is not in fact a true fir and the latter isn't a pine—or an umbrella, for that matter.

The names *pine*, *fir*, and *cedar* are each attached to a genus with a specific set of defined characteristics. Any tree that fits botanically within that genus is considered a "true" species, such that a true cedar is in the genus *Cedrus*, a true fir is in the genus *Abies*, and a true pine is the genus *Pinus*. Trees that are commonly called one thing but botanically are another are usually referred to as "false" species. In other words, a false cedar is a tree we call a cedar but that is in fact not in the genus *Cedrus*.

SCALE-LEAF AND AWL-LEAF CONIFERS

If you are thinking, "The scale-leaf conifers are confusing," I am with you in solidarity. The cypresses (false or otherwise) confound almost everyone. They not only look strikingly similar but are also the victims of taxonomic reorganization and misleading common names. So, yeah, they're confusing to say the least. All trees in the Pacific Northwest with scale- or awl-like leaves are in the cypress family, Cupressaceae, except for pehuén (or monkey puzzle, which is in Araucariaceae). I've included pehuén in this section because its leaves look more like scales than needles; I also figured it would stand out well enough on its own anyway, so it wouldn't hurt to include it.

The genera of the cypress family are broken down by their seed cone morphology. So, while their leaves may look similar across genera, their cones differentiate them (though you'll be able to tell them apart by their foliage in no time, I promise). They also tend to be large, with reddish, fibrous bark, ranging in texture from hard and woody to strappy and stringy.

The best way to tell these species apart is to look at the foliage first, then find a cone. The foliage (which is also the twig) of the scale-leaf conifers usually fits snugly into one of two camps: flattened along a single plane or not. If it's flattened, flip the twig upside down and look at the stomatal patterns on the underside. Each of the common species in the Northwest has its own pattern created by the leaf arrangement, leaf shape, and stomatal lines. Pair this pattern with the cones attached to the tree, and you'll have your species. It's just as easy to look at the cone first, then the leaf patterns, but just be sure the cone is from your tree—be wary of cones on the ground. If the twigs are not flattened into one plane, you'll need to take a closer look at how the smallest twigs grow from their stems. The junipers (*Juniperus*) and the true cypresses (*Cupressus* and *Hesperocyparis*) both have foliage that isn't flattened, but each has unique arrangements and leaf traits that set them apart.

If your tree has awl-like foliage, you can narrow down the possible species to a small handful of genera, with about six or seven common species between them in the Northwest. If you find this foliage type, it's often best to simply compare the traits of the leaves. They're all evergreen, but they each have particular attributes that can easily set one genus apart from another (and some genera have only one species to worry about anyway). As always, the next best step is to look for the cone. If you can find a cone, you've made your identification because each genus differs from the others in its cone substantially. Between the leaf traits and the cones, you're likely to sort out all but

These are the cones of the cypress family trees that are most common in our region. I've shown them all here together so you can get a sense of their size and shape relative to each other. Starting in the upper left, and moving top to bottom in four columns: Juniper "berries", Hiba-cypress, cryptomeria; second: Chinese-arborvitae, western redcedar, incense-cedar; third: Nootka cypress, hinoki cypress; fourth: true cypresses, sawara-cypress, Port Orford-cedar.

the few juniper species in the region, but I'll get to those a bit later.

I encourage you to take a look at the species comparisons (above and in table 1 on the next page) to familiarize yourself with the leaf and cone traits of the Northwest's common genera and species. If you're looking to really internalize them, I encourage you to draw the foliage and cones at something like three times their actual size. Observing them in such depth will help you notice specific differences, and you'll remember all those details far better than if you just stare at them. Drawing skills are not important; this exercise is meant to help you examine the leaves intentionally.

TABLE 1. QUICK GUIDE TO SCALE-LEAF AND AWL-LEAF CONIFERS

DESCRIPTION	FOLIAGE	CONE
Western redcedar Foliage flattened, underside has "butterfly" pattern in stomatal bloom; cones resemble rosebuds; arborvitae is similar, except no stomatal bloom below		
Hiba-cypress Foliage flattened, bright green, shiny, appears almost like armor; contrast between stomatal bloom and leaves on underside is bright; cones small, with wide scales		
Chinese-arborvitae Leaves small, roundish but in flatted splays, dark green; often an upright variety is found with vertically oriented leaves; cones whitish before opening, with hooked tips		
Port Orford-cedar Foliage flattened, underside has sharp X pattern in stomatal bloom; cones round, with small, blunted point on each scale; dense canopy		
Hinoki-cypress Foliage flattened, underside has I or Y pattern in stomatal bloom; cones round with no points on scales, bigger than sawara-cypress		
Sawara-cypress Foliage flattened but more rounded; stomatal bloom where pointed scales slightly overlap, a sloppy X sometimes; cones round, smallest of *Chamaecyparis*		

DESCRIPTION	FOLIAGE	CONE
Incense-cedar Foliage flattened, with no obvious stomatal bloom; scales like stacked cocktail glasses; cones look like a duck's bill with its tongue out		
Nootka cypress Foliage flattened and pendulous, but individual twigs rounded with no stomatal bloom; cones rounded, with distinctive point on each scale		
Leyland cypress Foliage like Nootka's, but decidedly not pendulous, often with stomatal bloom around scale edges; cones like Nootka's but much larger		
True cypresses Foliage rounded, ropelike, grows in multiple planes, scales don't overlap; cones greater than 1 in. (2.5 cm), rounded		
Junipers Foliage either like cypresses' or sharply awl-like, often on same twig; cones rounded, blue, juicy		
Japanese cryptomeria Foliage always awl-like, waxy, sharply pointed, usually lifted off twig; cones rounded, with many spikes on each scale		
Giant sequoia Foliage stringy, awl-like, bluish, sharply pointed, much smaller than cryptomeria's; cones the size of goose eggs, with thick, peltate scales		

CLOCKWISE FROM TOP LEFT: *Hinoki-cypress bark, arborvitae foliage, arborvitae cones, hiba-cypress foliage*

FALSE CEDARS AND FALSE CYPRESSES

Cupressaceae

Strictly speaking, false cedars and false cypresses aren't one group of trees, but rather a loose collection of cypress family (Cupressaceae) genera and species that all look fairly similar and so have been lumped together colloquially. The similarities are mainly in their scale-like leaves, so lumping them together by leaf trait makes a lot of sense. Botanically, however, we know that they are each unique because their cones are all quite different from each other.

Depending on how you lump or split, there are five to seven different genera with species that we call false cedars or false cypresses. You can usually tell them apart from the true cypresses (in *Cupressus* or *Hesperocyparis*) by their mostly flattened leaves that tend to slightly overlap one another. True cypress leaves abut one another rather than overlap, and true cedars (*Cedrus*) have needle-like leaves. Sometimes junipers (*Juniperus*) are also called cedars, but they have blue, succulent cones and either scale-like leaves (like the true cypresses) or sharp, awl-like leaves.

Western Redcedar
Thuja plicata

This species is by far the most common and widespread of the native false cedars west of the Cascades. It's planted in parks, yards, and along streets and grows natively in all sorts of places, but it favors wetter areas. If you see a false cedar growing in a yard or natural area west of the Cascades, it is most likely this species.

The cultivar 'Green Giant' is very upright, with bright green foliage. It is planted often and is probably a hybrid with *T. standishii*. Its "butterfly" pattern and cone set this tree apart, along with its size: it's by far the largest of our false cedars by almost all measures.

Large evergreen, often up to 70 feet (21 meters) tall in the landscape, but well beyond 100 feet (30 meters) in habitat; wide, pyramidal canopy with drooping tip and somewhat pendulous, feathery textured foliage on upswept branches that often turn up into leaders of their own. **BARK:** Reddish-orange, thick, fibrous, with strappy, shallow furrows. **TWIGS & LEAVES:** Small, appressed to twigs in alternating, opposite pairs; lateral leaves much larger than facial pair, with distinctive *"butterfly" pattern in white stomatal lines below*; foliage very fern-like in mostly flattened sprays, except on newest foliage in direct sunlight, which tends to be longer and rounder. **CONES:** Up to ½ inch (1.3 centimeters) long, elongated, and sitting upright on foliage sprays in dense groups; *resemble tiny rosebuds*, appearing green in winter but turning brown by fall; scales are oppositely arranged, overlapping. **SPECIES REMARKS:** Although foliage may turn bronze in winter, it is not dying.

bright green leaves and thick bands of white stomatal bloom on the underside of the leaf is striking to the point of being quite mesmerizing. This alone is characteristic of the species—no other trees you'll find even come close.

Chinese-arborvitae (*Platycladus orientalis*) most closely resembles the more common arborvitae (*Thuja occidentalis*), but instead of growing as a tight, upright column, it becomes a larger, rounded tree. It's common throughout the region, especially farther east. With close inspection, you'll notice the leaves are smaller, giving the foliage a daintier appearance overall. The cones are a primary clue: they start out looking like a larger, whitish-blue western redcedar cone, but each scale has an extended, hooked tip. When mature, they turn brown and open, reminding me of a wooden, spiky jack (as in the children's game, Jacks).

Similar Species

Arborvitae (*Thuja occidentalis*), also called northern whitecedar, is the classic hedge tree found along property lines across the country. Is it a tree? Most of the time, I'd say no. But it appears outside of hedges every so often and can look similar enough to its close relative, our native western redcedar, to warrant mentioning. You can usually guess the species by the way its cultivars look in the Northwest: tall and skinny and usually packed in rows. The leaves are a lighter green than western redcedar's, and the cones are smaller and plumper (see p. 64). Arborvitae's leaves are also more flattened and lack any stomatal boom below, which is a surefire way to tell it apart (also, p. 64).

Hiba-cypress (*Thujopsis dolabrata*) looks like western redcedar at first glance, but upon closer inspection, you'll notice the foliage is much more rigid and glossy dark green on top, almost like armored scales in threes (see p. 64). As in the top right photo, the contrast between the

Port Orford-Cedar
Chamaecyparis lawsoniana

Also called Lawson-cypress, it's a native endemic species of the coastal forests of the Siskiyou and Klamath Mountains. Port Orford-cedar is heavily planted as a landscape tree throughout the Northwest. It is almost only found in landscape plantings and does not grow wild outside its native area.

Large evergreen up to 60 feet (18 meters) tall, usually with single stem; foliage often semi-bluish, densely layered with shaggy, drooping appearance; usually not more than 20 feet (6 meters) in spread. **BARK:** Reddish-brown, fibrous-looking but firm and woody to the touch; develops large plates separated by furrows and often silvery with age or lichen; broken up by old branch scars. **TWIGS & LEAVES:** Flattened sprays of scale-like foliage; feathery appearance with long, skinny twigs covered in tiny, oppositely arranged and alternating scales; underside of leaves show an X *pattern in white stomatal bloom*, distinctive of this species. **CONES:** Small, rounded (globose), ⅓ inch (8 millimeters) in diameter, with 8 peltate scales, each with a tiny, blunted point in the middle; at first light bluish-green, covered in waxy bloom, maturing to a medium brown in fall; often found in big groupings on foliage.

SPECIES REMARKS: Port Orford-cedars are susceptible to the root fungus *Phytophthora lateralis*, which has infected many members of the native and ornamental populations, leaving fully dead, brown trees across the landscape. Hundreds of cultivars exist, but the X pattern on the underside of the leaves along with the comparatively large, round cones and hard bark are consistent identification keys for the larger trees.

Hinoki-Cypress
Chamaecyparis obtusa

This native of Japan is found mostly as a small tree in gardens west of the Cascades, very rarely as a street tree, and never in natural areas, as it does not seed itself in. You'll likely find the cultivars 'Aurea' and 'Compacta' most often, but the normal species is present.

Medium evergreen, usually less than 50 feet (15 meters) tall, with single stem and rounded, pillowy form; most often appear as small cultivars with rounded, layered foliage sprays. **BARK:** Reddish-brown, fibrous, strappy; often splitting into many shallow fissures. **TWIGS & LEAVES:** Flattened, spreading sprays of scale-like foliage; twigs tend to branch and splay outward, creating a curved fan shape; foliage *undersides have distinct* I or Y *pattern* in stomatal bloom, resembling a wrench or a cartoonish bone (see table 1).

CONES: Small, rounded, ⅓ to ⅜ inch (8–10 millimeters) in diameter with 8 peltate scales that *lack a distinctive central point*; orange-brown, opening in fall; largest of cones in *Chamaecyparis*.

SPECIES REMARKS: You'll almost always find a few dwarf cultivars with attractive, layered but rounded habits and bright green, almost yellow foliage. Hinoki-cypress's canopy is dense, but not as thick nor as droopy as Port Orford-cedar's. The I or Y pattern on the underside of the leaves; the fissured bark; and the very rounded, fan-shaped sprays of foliage are the keys to this species.

Sawara-Cypress
Chamaecyparis pisifera

Another Japanese species, sawara-cypress is planted west of the Cascades about as often as hinoki-cypress. The normal species gets large, but the often-planted, strappy-leafed cultivars stay much smaller. It does not seed itself in, so you won't find it in natural areas.

Large evergreen up to 70 feet (21 meters) tall, with single stem; shaggy, feathery appearance, lacier than others in genus; popular cultivars have especially wispy foliage. **BARK:** Reddish-brown, thin, fibrous, often with stringy fibers that are easily pulled off; doesn't tend to develop rough fissures. **TWIGS & LEAVES:** Small, overlapping, appressed to twigs in alternating pairs with pointed tips slightly reflexed, almost awl-like but not quite; slightly rounded; white stomatal bloom underneath where scales overlap, appearing like a shadow under the scale tip or sometimes like an X, but sloppily drawn; foliage is flattened on normal species, but stringier on some cultivars. **CONES:** Smallest of *Chamaecyparis*, rounded (globose), ⅙ to ¼ inch (4–6 millimeters) in diameter with 6–8 peltate scales; orange-brown, solitary, on short stalks at the end of twigs.

SPECIES REMARKS: The cultivars 'Filifera,' 'Squarrosa,' and 'Boulevard' are commonly planted. The latter two have only juvenile foliage, which is distinctly bluish, awl-shaped, and not very rigid (see photo, bottom right); by appearance alone, they could be confused with Japanese cryptomeria. 'Filifera' has longer, wispy foliage with a canopy described quite accurately by Patrick Breen of the Oregon State Landscape Plants Database as a "droopy mound." Tell it apart from others by its tiny cones and strappy, somewhat pokey foliage.

Incense-Cedar
Calocedrus decurrens

Another of our native false cedars, this one is found on the drier side of the Cascades and Siskiyous of Oregon south through California. It's planted far and wide on the west side of the Cascades as a drought-tolerant tree, but less so in the eastern portion of the Northwest.

Large evergreen up to about 70 feet (21 meters) tall in the landscape, bigger in habitat; single stemmed with dense, bright canopy and foliage often held vertically. **BARK:** Brown, scaly, with reddish furrows when young, often with obvious marks where old limbs were; with age, it becomes thick, with deep furrows between hard plates. **TWIGS & LEAVES:** Flattened, spreading, distinctly jointed sprays; elongated, light green leaves with *no stomatal bloom below* but lighter green borders; scales resemble long-stemmed cocktail glasses stacked inside one another. **CONES:** Pendulous, almond shaped, light green when immature, turning light brown and opening in fall; just 2 large scales with hooked tips reflex back away from flat central axis, appearing like a duck's bill yawning with its tongue out. **SPECIES REMARKS:** The lack of stomatal bloom on the long, light green, wine-glass leaves and the fabulous cones are the best keys for this species, unique among any of our conifers. The bark is also very hard and rigid, which can separate it further.

CLOCKWISE FROM TOP LEFT: *Nootka cypress bark; Monterey cypress cone; Monterey cypress form; Modoc cypress leaves*

TRUE CYPRESSES

Cupressus and Hesperocyparis

Are you a **lumper** or a **splitter**? Taxonomists are often somewhat cheekily put into these two camps: those who broadly lump populations together, and those who insist they be split into discrete **taxa**. It boils down to where you draw the line and what traits you think indicate unique, important differences. If two trees are almost identical but one has pubescent leaves and the other does not, should each qualify as its own distinct species, subspecies, or variety? Or is there simply natural variation within one true population? Taxonomists have created heaps of studies over hundreds of years, and their debates often result in some new rearrangement of species. At best, these decisions give us a more accurate accounting of species. At worst, they seem arbitrarily academic and force us to relearn a new batch of scientific names.

Cypresses have been the subject of this intense taxonomic scrutiny for the last several decades. Once upon a time, all the "true" cypresses were in a single genus, *Cupressus*. Then, during the first few decades of the twenty-first century, taxonomists segregated the genus into *Cupressus* (made up of the species from Eurasia and Africa) and *Hesperocyparis* (made up of species from North and Central America). To add to this, there is uncertainty within *Hesperocyparis* itself—namely, whether there are several independent species across the US Southwest and Mexico, or just one with several isolated populations that should be considered varieties of Arizona cypress, *Hesperocyparis arizonica*.

This discourse will likely continue for some time, but for the purposes of this guide, the true cypresses will appear in both *Cupressus* and *Hesperocyparis*. Trees in both genera have rounded, scale-like foliage (almost never awl-like) and spherical cones. Incidentally, I'm also on team "Nootka Cypress Is a True Cypress," so I've omitted the hyphens for it and Leyland cypress in quiet protest. I'll let you investigate that taxonomic quagmire independently so that you can make up your own mind.

Nootka Cypress

Callitropsis nootkatensis
Syn. *Chamaecyparis nootkatensis, Cupressus nootkatensis*

Also known as Alaska yellow-cedar, this high-elevation species grows in the mountains in the western portion of the Pacific Northwest. Though native, in the lower-elevation landscapes, it doesn't seed itself in, but it is planted by the dozens for its graceful habit, thin size, and slower growth.

Medium evergreen, usually up to about 40 feet (12 meters) tall in the landscape, twice that in habitat; single stemmed with sparse canopy and *drooping foliage* (intensely so with common cultivars). **BARK:** Grayish-brown when young, tending to exfoliate in irregular patches; with age, becomes a weathered gray, with long fibrous splits. **TWIGS & LEAVES:** Pendulous, flattened sprays, with individual twigs slightly rounded; dark green with bluish tinge to grayish-green; small leaves with *underside that lacks white stomatal lines*, but an X-like pattern in light green where scales overlap may still appear; scale tips sharp when brushed up the twig; all scales are nearly the same size; emit intense, malodorous smell when bruised. **CONES:** Globose, about ¼ to ½ inch (6–12 millimeters) in diameter with usually 4 (sometimes 6) scales, each with a *distinct pointed projection*.

SPECIES REMARKS: Though it's already a pendulous tree, an even more pendulous cultivar is mostly planted, making for a weepy canopy of foliage that drapes off the limbs; no other species has a similar form. Look for the lack of stomatal bloom and the spherical cones with sharp points on the scales for this species.

Leyland Cypress
× *Hesperotropis leylandii*

Leyland cypress is a hybrid between Nootka cypress and Monterey cypress, and it's planted by the dozens in the landscape. It's a cheap, large, upright evergreen that grows unsustainably fast, often used in hedges. There are too many synonyms to name, but they all end with *leylandii*.

Medium evergreen, single stemmed, usually up to about 60 feet (18 meters) tall, with very upright, dense growth habit; looks like an explosion (to me), with all limbs growing upward and out. **BARK:** Brown, strappy in appearance, but not fibrous. **TWIGS & LEAVES:** Dark green to bluish-green, arranged in opposite pairs; closely resembles Nootka cypress, but foliage grows up and outward and isn't flattened, which is similar to Monterey cypress; lateral and facial scales are approximately the same size; does not emit distasteful odor. **CONES:** Usually absent but appear every now and then; resemble Nootka cypress's, but larger; round, ½ to ¾ inch (1.2–2 centimeters) in diameter, with slight points on the scales.

SPECIES REMARKS: This is the new-age hedge tree because it grows so quickly: after you plant it, you'll have a 10-foot (3-meter) hedge before you know it; look away for a second, and it'll be 30 feet tall and 20 feet wide. It's adaptable, so it's found across the Northwest, and can be identified by the dark green, upright foliage and placement along a property line.

Mediterranean Cypress

Cupressus sempervirens

Also known as Italian cypress, a classic Mediterranean countryside tree, this is our first "true" cypress. Natively, there are two forms of this species, one that spreads to create a wider canopy and another that grows upright like a pillar. I've only ever seen the upright pillar planted in the Northwest, so that's what is described here.

Small evergreen, single stemmed, usually not taller than 25 feet (8 meters) or wider than 3 feet (1 meter) with intensely columnar habit; dense growth with wisps of foliage often reaching out of older trees. **BARK:** Gray, fibrous, often obscured by foliage. **TWIGS & LEAVES:** Small, dark green, diamond shaped, arranged in opposite, alternating pairs that seamlessly abut one another; all scales are about the same size, tightly appressed to the twig without a sharp point; twigs rounded, and foliage not flattened; older twigs brownish-beige. **CONES:** Large compared to other scale-leaf trees, 1 inch (2.5 centimeters) in diameter, gray to brown, egg shaped, smooth but for small, mostly blunt points on ends of scales, like an armored egg sac.

SPECIES REMARKS: The form should differentiate this species from most others save for arborvitae and the pillar cultivar of Monterey cypress (see p. 76). To separate them, look for the finer-textured, rounded foliage that appears slightly stringier than Monterey's (arborvitae is flattened). The cone is rounded but usually oblong, while other cypresses' cones are more spherical.

Arizona Cypress

Hesperocyparis arizonica

This native of southwest North America is being planted more and more often in the Pacific Northwest as a drought-tolerant and attractive evergreen. Arizona cypress does not grow wild in our region and does not seed itself in, so you'll find it only where it has been planted.

terranean cypress's) and stouter, whitish-blue foliage. The native **Modoc cypress** (*H. bakeri*) looks similar and is sparingly planted in the Pacific Northwest but has greener, more delicate twigs with more obvious resin dots. Its foliage is also decussate but more closely spaced, so it appears slightly denser, but also finer (see below and p. 71).

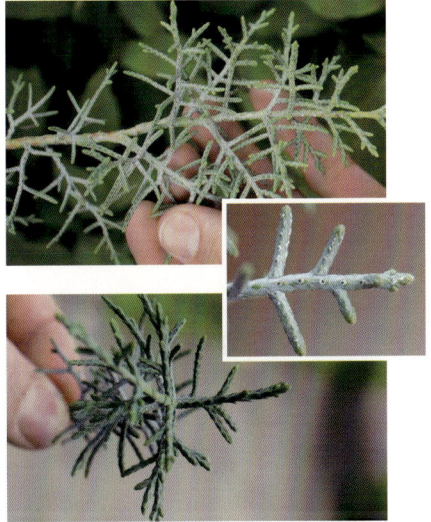

Medium evergreen up to 40 feet (12 meters) tall, single stemmed with open, pyramidal, airy but regimented canopy; intensely bluish cultivars are most often planted, giving it an icy look. **BARK:** Smooth, reddish-brown, with tones of yellow and olive green; outer layers flake off attractively; can become furrowed with age. **TWIGS & LEAVES:** Small, diamond-shaped scales tightly appressed to fairly stout twigs in alternating pairs, creating a rounded (really, squared) twig; *usually an obvious resin dot on leaf surface*, but not always; *light green with bluish cast due to waxy bloom*; foliage twigs irregularly spaced and not flattened; twigs grow in *decussate arrangement* (if you look down a new twig, it'll look like a +). **CONES:** Rounded, up to 1 inch (2.5 centimeters) in diameter, usually glaucous at first, with 3–4 pairs of scales. **SPECIES REMARKS:** Look for the big, spherical cones (not egg shaped like Medi-

BELOW: *Modoc cypress form and decussate leaf orientation*

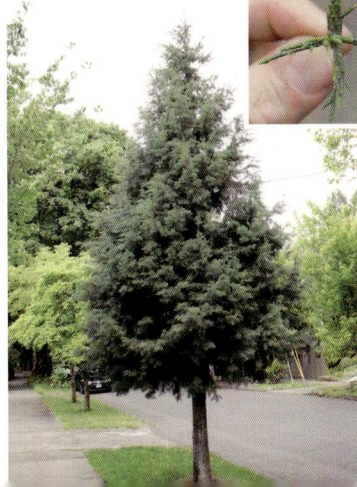

Similar Species

Monterey cypress (*Hesperocyparis macrocarpa*) is commonly found planted along the coast of Oregon. Elsewhere west of the Cascades, you'll more often find one of its yellow-leafed cultivars (see below, top right) or an upright form suspiciously similar to Mediterranean cypress. The natural species form is a large tree up to 80 feet (24 meters) tall with a dense, wide-spreading canopy and often multiple main trunks, easily far larger than all our other cypresses.

Its small, tightly appressed leaves are dark green without any glaucous bloom and *no resin dots*. The mature bark appears strappy and shaggy but is mostly hard, with long vertical fissures. The twigs are bright reddish brown, which will help you discern it from Mediterranean cypress. The cones are round like other cypresses, about 1 inch (2.5 centimeters) in diameter, gray, and smooth, but bumpy and not glaucous.

CLOCKWISE FROM TOP LEFT: *Both scale-like and awl-like leaves on the same twig; Utah juniper leaves; western juniper leaves and fruit*

JUNIPERS

Juniperus

Two traits help you easily recognize the junipers (family Cupressaceae): their fleshy cones, resembling berries, and their dimorphic leaf growth, either scale-like or awl-like, often on the same twigs. Young trees, or trees that have been damaged by grazing animals or hedge-trimmer-wielding humans, grow sharply pointed, almost needle-like leaves as a defensive measure. More mature, undamaged sprigs grow entirely scale-like leaves, tightly appressed to the twig. This is a great identification characteristic, as no other genera will have two different, contrasting leaf types right next to each other.

When it comes to their cones, junipers are one of the few conifer genera that have adapted their cones for dispersal via animals, namely birds. Their cones are small, round, and often bluish and are covered with waxy bloom, but instead of drying out when they are mature, they become succulent, which attracts birds who consume them readily. If you see a bluish, fleshy cone, you know you've got a juniper.

There are several species native to the Pacific Northwest, but only a few are commonly found in towns, especially on the east side of the Cascades. They can be tricky to tell apart, so context and location are often helpful. *Conifers of the Pacific Slope* by Michael Kauffmann is a great guide for identifying conifers in their native habitats.

Chinese Juniper
Juniperus chinensis

The most common juniper planted on the west side of the Cascades, Chinese juniper mostly appears as a low-growing shrub, but often enough it'll be an upright tree. The cultivar 'Torulosa', known as the Hollywood juniper, is the most common tree type.

⅓ inch (8 millimeters) long, sharp, slender, awl-like, *in whorls of 3 or pairs;* adult leaves small, scale-like, diamond shaped, and in alternating pairs; *lacks resin dot on outside of leaf.* **CONES:** Globose, small, no more than ½ inch (12 millimeters) in diameter; bluish when young, covered with whitish bloom before maturing to violet-brown; dioecious, so female clones may have cones but males do not.

Small evergreen, usually no more than 30 feet (9 meters) tall, single stemmed with upright, pyramidal growth; species variety has rounded crown, while 'Torulosa' has irregularly branched canopy with outstretched limbs. **BARK:** Brownish-gray with fibrous, exfoliating strips broken into shallow furrows with age; stems appear ripply or braided. **TWIGS & LEAVES:** Green to greenish-gray, of 2 kinds: juvenile leaves

SPECIES REMARKS: The leaves of Chinese juniper don't have distinctive resin dots (unlike western juniper), and the foliage tends to be slender and green without much whitish bloom (unlike Rocky Mountain juniper). There are numerous cultivars of this species, and potentially a hybrid with *J. sabina,* so the variety in forms is substantial.

Western Juniper
Juniperus occidentalis

East of the Cascades, but not quite to the Rocky Mountains, this is the most common juniper. It's rarely planted but can be found growing all over the place, distributed by birds. West of the Cascades, it's an uncommon landscape tree.

Medium evergreen up to 30 feet (9 meters) tall, sometimes taller, with upright, pyramidal growth form when young, irregularly spreading with age. **BARK:** Light brown, flaky, and exfoliating on young trees, becoming gray, shaggy, and fibrous with age. **TWIGS & LEAVES:** Gray-green to dark green, usually scale-like (adult leaves), diamond shaped, appressed to twig *with resin gland on each leaf*, often with dried resin; usually *arranged in whorls of 3* (ternate), thus twigs branch out in 3 main directions (as opposed to decussate growth with 4, as in cypresses); juvenile foliage awl-like and very sharply pointed, often in pairs on new, vigorous shoots. **CONES:** Globose, small, bluish-black, covered in light blue waxy bloom; mostly smooth with a tiny point on some scales; mostly dioecious, but often with a few seed cones on trees that otherwise only produce pollen.

SPECIES REMARKS: Commonly invading unburned land east of the Cascades, western juniper has a bad reputation for being an aggressive colonizer. It's the primary wild juniper you'll see in Oregon and Washington east of the Cascades and in southern Idaho; aside from this context, the scale leaves in whorls of three with obvious resin dots and the cone that remains bluish black are excellent identifiers.

Rocky Mountain Juniper
Juniperus scopulorum

This is the primary native species in and east of the Rocky Mountains, found sparingly in Oregon and Washington, but more commonly in the north and east of the Pacific Northwest. Several upright, blue-colored cultivars are planted in the landscape throughout the region.

Small evergreen, around 30 feet (9 meters) tall, with rounded crown; species foliage often somewhat flaccid, hanging shaggily off branches; blue cultivars very upright. **BARK:** Reddish-brown to somewhat gray, fibrous, exfoliating in thin strips. **TWIGS & LEAVES:** Light to dark green, variably with glaucous covering or not; scale leaves are diamond shaped and do not (or just barely) overlap, with slightly reflexed tips; each leaf has a resin duct but, in contrast to western juniper, no resin drops, and are mostly in alternating pairs, not whorls of 3; awl-like leaves fairly rare. **CONES:** Dioecious, *pollen cones and seed cones found on separate trees*; small, ¼ to ⅓ inch (6–8 millimeters) in diameter, intensely bluish with white bloom; matures to a darker blackish-purple. **SPECIES REMARKS:** In the landscape, the upright, bluish cultivars may appear like true cypresses; the foliage is far coarser on the junipers and more bluish (which also separates Rocky Mountain juniper from Chinese juniper, which is mostly green). **Utah juniper** (*J. osteosperma*) is rarely found in the southeast of the Pacific Northwest, not often in towns and cities. Its leaves are more yellow green (see p. 77), its cones larger, and its crown broader and less pyramidal, with gray bark. It's also monoecious, so both types of cones are on the same tree.

BELOW: *Utah juniper fruit and bark*

Japanese Cryptomeria
Cryptomeria japonica

A native forest tree of Japan (also called sugi or Japanese-cedar), cryptomeria is planted on the west side of the Cascades as a landscape tree. You'll find it planted as a yard or park tree, rarely as a street tree, and you won't find it in natural areas.

Medium evergreen up to 60 feet (18 meters) tall, with pyramidal form; single stemmed with many branches; dense, bright green canopy. **BARK:** Red, fibrous, often exfoliating in long, stringy strips; becomes more furrowed with age. **TWIGS & LEAVES:** Bright green, covered with spirally arranged, awl-like leaves, each ¼ to ¾ inch (6–19 millimeters) long; leaves arch toward end of twig, with sharply pointed tips and a base that wraps around twig; very waxy and firm.

CONES: More or less rounded, but with 20–30 scales, each with a few sharp, bristly tips; looks like a spiky, brown fireball.

SPECIES REMARKS: There are many cultivars of this species, and all but 'Elegans' have leaves like the species. 'Elegans' cryptomeria differs substantially by having longer, more widely spaced leaves that point out perpendicular to the twig, like needles (see below, top right). This cultivar also turns entirely purplish bronze in the wintertime before turning back to dark green in warmer weather. The cone may be present on most cultivars, and it is the best identification characteristic, as no others look remotely similar. The foliage is also unique, but be careful not to confuse it with the Boulevard sawara-cypress, which has softer foliage.

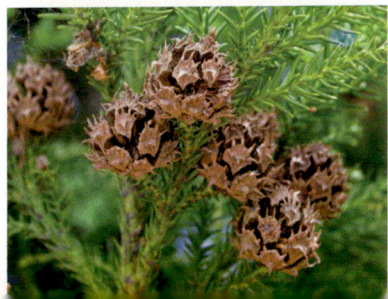

81

Giant Sequoia
Sequoiadendron giganteum

Often conflated with coast redwood (*Sequoia sempervirens*), giant sequoia grows natively in the High Sierra of California, not near the coast at all. It's planted widely in the Northwest in parks, yards, and streets, but it is not found in natural areas unless it's been planted. Historically, the redwoods had their own family (Taxodiaceae), but they are now all in Cupressaceae.

Giant evergreen up to 100 feet (30 meters) tall in the landscape, triple that in habitat; extremely pyramidal growth form that becomes rounded with age; crown dense, very uniform, with single stem that gains diameter very quickly. **BARK:** Red, thick, fibrous, with deep furrows as it ages; base expands quickly, so bark thickens quickly. **TWIGS & LEAVES:** Dark bluish-green, small, awl-like (not linear or needle-like), overlapping along strappy twigs; leaves point forward in groups of 3; sharply pointed and painful if rubbed backward. **CONES:** Large, egg shaped, growing continuously until they fall off, usually maxing out at around 3.5 inches (9 centimeters) long; cone scales diamond shaped and wrinkly, abut one another and do not overlap.

SPECIES REMARKS: Though closely related to coast redwood, giant sequoia has only awl-like leaves, never linear or needle-like. Giant sequoia also has much larger cones. It maintains its pyramidal shape until it's very old, which is a great identification characteristic from afar. The leaves, cones, and overall shape will set it apart from all other trees in the Pacific Northwest. A fun cultivar is the very silly Pendulum.

BELOW: *Pendulum cultivar*

Pehuén
Araucaria araucana

Found in the west of our region, this tree (also known as monkey puzzle) hardly needs a description at all due to its unique appearance. As its leaves defy easy categorization, morphologically it fits best here, somewhere between raised scales and flattened needles. It's native to the southern tip of South America and has regained the appeal it once had a century ago as a very curious specimen tree in parks and yards. **Family: Araucariaceae**

Large evergreen up to 60 feet (18 meters) tall in the landscape, with single stem; when young, very pyramidal, open habit with widely spaced, upward-arching branches covered with triangular leaves from near the base to the topmost limbs; develops rounded top with age, with only living branches and foliage at top. **BARK:** Gray with old leaves sticking out for years; with age, loses leaves and develops rippled appearance with pronounced branch scars. **TWIGS & LEAVES:** Large, dark green, triangular leaves cover twigs densely; 1–2 inches (2.5–5 centimeters) long, flattened, appearing like broad leaves; sharply pointed and very stiff. **CONES:** Mostly dioecious, with both large pollen and seed cones very obvious on respective trees; pollen cones 6 inches (15 centimeters) long, cylindrical, appearing at the ends of branches in clusters of 1–5, densely bristly; seed cone solitary, sitting upright on branches, with wide base and dense, bristly scales; turns brown and falls apart scale by scale at maturity. **SPECIES REMARKS:** When you see a pehuén, you'll know it. The common name I'm using is what it's called in its native area rather than the fun, but ultimately nonsensical, "monkey puzzle," which is a relic of colonial Victorian England.

NEEDLE-LEAF CONIFERS

Conifers with needle-like leaves are by far the most common type of conifer and are familiar to most people in our region. These are the classic pines, firs, and spruces, of which we have several native species, and even more planted as ornamentals. You can differentiate between genera and some species by the needle arrangement, their number, and ultimately their unique shapes and traits. First, count how many needles are growing from a single point: if it's just a single needle attached to a particular spot along the twig, then it's singly borne and listed in the first half of this section. Next, look at the attributes of the leaves, including their size, orientation, sharpness, and shape; these together should tell you the genus with little trouble. Finally, find the cone. Remember that each species has its own unique cone, and when it's paired with a tree's foliage

CLOCKWISE FROM LEFT: *Author with a cedar; pine needle bunch; spruce leaves and cones*

Dawn redwood grove

traits, you will be able to nail it down with confidence (see table 2).

If more than one needle appears to be growing out of the same point, then you'll need to figure out how many. If they are connected at the base in bundles of two, three, or five, then you've got yourself a species of pine, in the genus *Pinus*. No other genus has needles in these discrete bundles, so this arrangement confirms you are dealing with a true pine. Additionally, each pine species consistently has a certain number, so you can further break down the pines into the two-needle, three-needle, and five-needle pines. There is some variation in these groups, and some species may have one needle more or less than normal, so look at a few examples to get an accurate needle count. The quick guides in the pine section go through the different species in each group and their unique traits.

If many needles arise from what appears to be one spot, but they are not connected together at the base in bundles of two, three, or five, then you can narrow down your options to just a few potential genera near the end of this section. At this point, compare specific traits of the leaves, followed of course by finding a cone. Again, you could skip directly to the cone to determine the species, but organizing your process in a repeatable way will help you keep things straight when starting out.

TABLE 2. QUICK GUIDE TO NEEDLE-LEAF CONIFERS

DESCRIPTION	FOLIAGE	CONE
Redwoods Needles single, flattened, 2-ranked, with 2 obvious lines of stomatal bloom below, or light green, soft, deciduous; pendulous cones with peltate scales, up to 1 in. (2.5 cm) long		
Yews Needles single, flattened, often 2-ranked, with faint or no stomatal bloom below; "cones" look like red berries		
Hemlocks Needles single, flattened, 2-ranked, with 2 lines of stomatal bloom below, or diffuse on both sides and starlike; pendulous cones with rounded, woody scales, no bracts		
Douglas-fir Needles single, grow in all directions, 2 lines of stomatal bloom below, slightly flattened, not sharp; pendulous cones with rounded scales and exserted 3-pronged bract		
Spruces Needles single, stout, usually sharp, growing on short woody pegs that persist on old twigs; pendulous cones, scales rounded and woody or triangular and papery, no bracts		

DESCRIPTION	FOLIAGE	CONE
True firs Needles single, stout, but not sharp, with suction cup–like attachment to twig, usually pointed out or upward; upright, dehiscent, barrel-shaped cones on upper limbs		
Pines Needles in bundles of 2, 3, or 5; cones woody with obvious umbo at end of each scale, pendulous or pointing away from twigs		
True cedars Needles in clusters on short spur shoots, at least 15 per shoot, waxy, stiff, evergreen; cone upright, dehiscent, barrel-shaped; appear throughout canopy		
Larches Needles in clusters on short spur shoots, at least 15 per shoot, soft, light green, deciduous; cones upright, up to 2 in. (5 cm) long, do not fall apart, persist on twigs for several years		

CLOCKWISE FROM TOP LEFT: *Comparison of redwood cones; coast redwood form; dawn redwood leaves; coast redwood leaves*

REDWOODS

Redwoods are internationally famous, and for good reason: these striking, charismatic trees capture the imagination like few others can, growing to massive proportions surprisingly quickly. While none are widely native (the coast redwood barely grows natively in the far southwest corner of Oregon), they are commonly planted across our region, especially on the west side, even making it into some plantation forests in the Coast Ranges. East of the Cascades, you're less likely to see coast redwood in lieu of baldcypress or giant sequoia.

The redwood group was once considered its own family, Taxodiaceae, named after the genus of baldcypress (*Taxodium*, p. 91). But today they are well-established members of the cypress family, Cupressaceae, though they're still grouped together in closely related subfamilies. Five species in this group can be found throughout our region. You've already met two members, giant sequoia and cryptomeria, which both have awl-like leaves. The other three have more needle-like leaves, so they're described in this section. Despite their pine family-like appearances, their cones have peltate scales and their bark is red and fibrous—remember to look for those traits to set them apart.

Coast Redwood
Sequoia sempervirens

This is the renowned redwood that famously grows along the coast in Northern California (and technically very southern Oregon). It's been planted widely in the landscape west of the Cascades, and along with giant sequoia, it's sometimes planted as a crop tree in plantation forests, though it doesn't grow from seed outside its native range.

Huge evergreen, grows quickly up to 100 feet (30 meters) or more in the landscape, but like giant sequoia, easily triple that in habitat; very pyramidal form until old and large; branches appear flat, layered, and often point up at tips. **BARK:** Red, thick, fibrous, with deep furrows; harder than giant sequoia, with deeper furrows. **LEAVES:** Stiff, flattened, needle-like except on high, exposed twigs and fertile twigs, which are appressed (awl-like, similar to giant sequoia's); 2-ranked and glossy green above, 2 stomatal lines below; apex sharply pointed. **CONES:** Like giant sequoia's, but much smaller, only up to 1 inch (2.5 centimeters) long; scales peltate, irregularly spaced when opened, and very wrinkly.

SPECIES REMARKS: Just as with giant sequoia, coast redwood grows big fast, and the two are often confused. The smaller cone and two-ranked, needle-like leaves easily set the coast redwood apart. Don't get thrown off if you see awl-like leaves near the cones on the otherwise needle-leaf redwoods, though: giant sequoia doesn't ever have needle-like leaves.

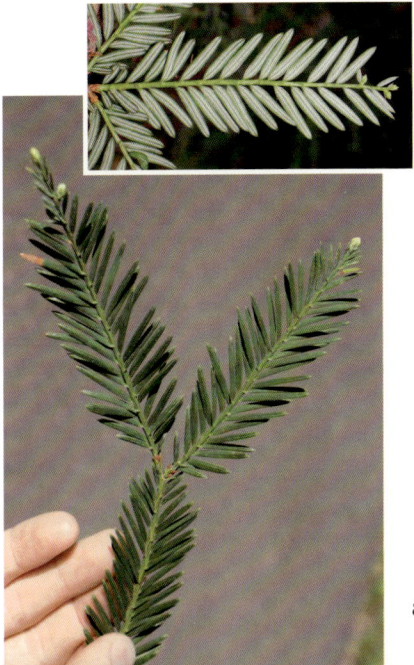

Dawn Redwood
Metasequoia glyptostroboides

A popular landscape and park tree, dawn redwood is a legendary tree lauded as a so-called living fossil due to remaining nearly unchanged morphologically for over 150 million years. You'll see this species in parks and along streets anywhere in the Pacific Northwest, but mostly west of the Cascades.

Large *deciduous* conifer up to 100 feet (30 meters) tall, with distinctly pyramidal form and single stem throughout its life (like giant sequoia); airy canopy with ascending branches. **BARK:** Reddish-orange to grayish-brown, often textured with flaky ridges; stem often develops bulbous, doughy ridges between and around branches, creating what looks like armpits below branches or braided ridges at the base of the trunk. **LEAVES:** Light green, soft, and always needle-like; 2-ranked and attached *oppositely* along twiglets, which are themselves *arranged oppositely along the main twigs*; needles turn a soft yellow-brown in fall, and *whole twiglet falls away with needles*. **CONES:** No more than 1 inch (2.5 centimeters) long, egg shaped but notably square in cross section due to oppositely arranged peltate scales that are offset by 90 degrees. Scales look like a pair of lips.

SPECIES REMARKS: This species is often confused with baldcypress but differs in its leaves, cones, and form. Baldcypress needles and twiglets are alternate, their cones are not square (more like owl pellets), and their form is not nearly as uniform and stately, but rather more rounded and messier.

Similar Species

Baldcypress (*Taxodium distichum*) is often confused with dawn redwood, and for good reason: they are closely related and look strikingly similar in most regards, down to their deciduous foliage. However, baldcypress is not nearly as common in the Pacific Northwest, so the odds are in favor of you finding a dawn redwood. Baldcypress has *alternately arranged leaves and twiglets* (in contrast to dawn redwood's opposite arrangement) that routinely lie nearly appressed to the twig, similar to a giant sequoia's, only deciduous; the cones are rounded, egg shaped, don't have nicely symmetrical scales, and either stay closed or break apart when mature; the bark is grayer and does not develop rounded ridges or armpits below the stems; and finally, all but a few cultivars develop woody growths from their roots, called "knees," that serve as gas-exchange organs in their native swampy habitats. Any of these traits would serve to help differentiate baldcypress from dawn redwood.

Japanese yew "berry," which is all but identical to the other species you may find in our region

YEWS

Taxus

Yews (family Taxaceae) are probably the strangest of our common conifers because they opted to essentially do away with their cones altogether. Adopting the way of the broadleaf trees, rather than depending on the wind to carry their seeds to distant places, they made an evolutionary bargain with animals—more specifically, birds. The "cone" has been reduced to just a few scales that fuse together around a single seed, become succulent and juicy, and turn red, creating a "berry" that we call an aril. If you see a conifer with a red berry, you'll know it's a yew without question.

One species is native to our region: **Pacific yew** (*Taxus brevifolia*), which grows wild in moist forests from the coast to Idaho and north to Southeast Alaska. However, this species is fairly uncommon in our urban areas, as it tends to be sensitive to disturbance and drought. More commonly you'll find **English yew** (*Taxus baccata*) or **hybrid yew** (*Taxus × media*) growing in the landscape, the latter a hybrid between English yew and **Japanese yew** (*Taxus cuspidata*). They all tend to grow as large shrubby trees in the landscape, so I'm only glancing over them here.

Pacific and English yews both have mostly two-ranked, needle-like leaves with pointed tips that are dark green on top and pale yellow green below, *distinctively lacking obvious stomatal lines*, which separate yew leaves from those of all other similar conifers (namely, coast redwood and grand fir). Pacific yew can be identified by the midrib on top of the leaf that runs only halfway down the needle from the base. While English and hybrid yews' leaves are similar, they have a ridged midrib from tip to base. English yew has slightly longer and wider leaves and develops sinuous ridges on its stem. Pacific yew tends to develop an undulating stem covered in multicolored, flaky bark ranging from pinks and reds to purples and browns. Both hybrid and English yew are mostly planted as very fastigiate cultivars, growing almost directly upward.

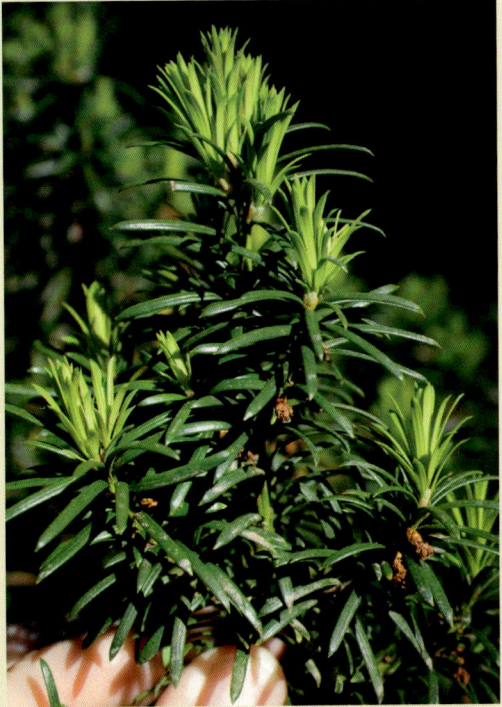

CLOCKWISE FROM TOP LEFT: *Pacific yew bark; Japanese yew leaf; Pacific yew leaf underside and leaf*

CLOCKWISE FROM TOP: *Comparison of our region's three hemlock cones with Douglas-fir. Clockwise from left: Mountain hemlock, western hemlock, Douglas-fir, eastern hemlock; eastern hemlock leaves; mountain hemlock leaves*

HEMLOCKS

Tsuga

Hemlocks (family Pinaceae) are one of the more common wild species in the Pacific Northwest, often dominating in late successional forests west of the Cascades and at high elevations in most mountain ranges here. In less wild landscapes, however, hemlocks tend to play a smaller role, ceding their dominance to species that fare better in more disturbed conditions. But that doesn't mean they're entirely absent.

You're likely to find just three species in our landscapes, and mostly only west of the Cascades. Eastern hemlock and an exceptionally skinny, blue cultivar of our native mountain hemlock are the two most common ornamental species. But our other native species, western hemlock, is planted as a landscape tree or more frequently seeds itself in, especially along the fringes of the landscape in areas closer to forests.

Hemlocks have singly borne needles that are neither stiff nor sharp to the touch, and their cones hang pendulously throughout the canopy. They often resemble spruces or Douglas-fir, but their needles are quite soft, and they have neither the woody pegs at their base that spruces have nor extended trident bracts on their cones like Douglas-fir.

Western Hemlock
Tsuga heterophylla

A tall native of the forests west of the Cascades, western hemlock is a common denizen of edge habitats and can often be found planted throughout the landscape. It's usually found with (and confused as) Douglas-fir, its much more common and distant cousin.

Large evergreen up to 100 feet (30 meters) tall in the landscape with pyramidal canopy; very feathery and soft textured; *tips of main leader and branches often nod over lazily.* **BARK:** Reddish-brown when young, becoming brownish-gray with age; rough textured, developing irregular, shallow furrows but not deep fissures. **LEAVES:** Needle-like, flattened, with short petiole and blunted tip, appearing mostly 2-ranked but with several needles irregularly angled away

from the stem, especially on vigorous shoots; some appearing upside down; *needles variable in length* from ¼ to ¾ inch (6–19 millimeters), lustrous green above, 2 stomatal lines below. **CONES:** Small, no more than 1 inch (2.5 centimeters) long, pendulous, often in great numbers on and below trees; best description is pretty dang cute.

SPECIES REMARKS: Look for the variable-length needles in more or less two ranks, the drooping branch and shoot tips, and the adorable cones. Mountain hemlock has uniform-length needles with no obvious stomatal bands and a starlike arrangement. **Eastern hemlock** (*T. canadensis*) has *consistently upside-down needles that run along the twigs,* whereas western hemlock's are more sporadic; otherwise they are extremely similar (see below). Eastern hemlock does not seed itself in, so any wild seedlings are surely western; eastern also has many more cultivars, so any strange-looking miniature or pendulous varieties are surely eastern.

BELOW: *Western hemlock leaf*
INSET: *Eastern hemlock leaf*

Mountain Hemlock

Tsuga mertensiana

Though a large native species that grows at high elevations, often forming the timberline, mountain hemlock is planted at lower elevations as a small ornamental landscape tree. It doesn't grow naturally from seed at low elevations, so you'll see it only where it has been planted.

Small evergreen, almost never more than 30 feet (9 meters) tall in the landscape, but easily triple that in habitat; tight, pyramidal to conical habit, usually with foliage to ground; pokey-looking texture caused by spiral foliage arrangement, but soft to the touch. **BARK:** Rough, almost scaly, brownish-gray in the landscape (often obscured by foliage); large, older specimens turn a richer red-brown. **LEAVES:** Uniformly up to ¾ inch (2 centimeters) long, slightly flattened, spirally arranged about twig but angled toward tip and concentrated on upper side; distinctly bluish-green with diffuse waxy bloom on both sides; *shorter shoots look like stars*, giving foliage sprays spiky, but not stiff, appearance. **CONES:** Cylindrical, 2–3 inches (5–8 centimeters) long, with rounded, slightly cupped scales that reflex entirely backward when fully dry, making the cone appear inside out.

SPECIES REMARKS: Short, bluish-gray cultivars are usually planted as small conifers in the landscape. Mountain hemlock can be confused with **Serbian spruce** (*Picea omorika*), which has a similar form but flattened needles (see table 3, p. 101); look for the pegs on old twigs indicating a spruce, as well as for two bright lines of stomatal bloom below; mountain hemlock lacks obvious lines, while they're quite clear on Serbian spruce. The spirally arranged, bluish foliage and larger cone set mountain hemlock apart from other hemlocks, and it lacks any bracts on the cones, which separates it from Douglas-fir.

Douglas-Fir
Pseudotsuga menziesii

The most common coni-fer on the west side of the Cascades, and not infre-quently found across the rest of the Pacific North-west, Douglas-fir (family Pinaceae) is nothing short of iconic. Botanically, it's not a true fir, hence the hyphen in the name. It is found nearly anywhere trees grow except where there is too much water or it's too dry.

Huge evergreen, easily growing over 150 feet (45 meters) tall, with pyramidal habit when young; becomes more irregularly triangular with age as the canopy gets larger; somewhat messy and shaggy tex-ture compared to hemlocks and true firs; great variability in form, from drooping twigs to sprightly, upright growth. **BARK:** Gray and smooth when young, with many resin blisters, but turns rich brown with age as cracks develop into deep furrows. **LEAVES:** Up to 1 inch (2.5 centimeters) long, nee-dle-like, slightly flattened, spirally arranged about the twigs, resembling a bottle-brush (see p. 98); light green on top with 2 obvious stomatal bands below; *buds are distinctly conical and sharply pointed.* **CONES:** Up to 4 inches (10 centimeters) long, cylindrical, pendulous with wide, rounded scales, each paired with a long, 3-pointed bract that looks like the hind end of a mouse in midleap under each scale (see p. 98).

SPECIES REMARKS: Usually the biggest tree in an area, Douglas-fir stands out from the crowd with its size. Only grand fir and western hemlock look similar and grow to a similar height west of the Cascades. How-ever, grand fir has upright cones and dis-tinctly two-ranked needles, and hemlocks (along with spruces and all other conifers

with pendulous cones in the Pacific Northwest) lack exserted bracts. There are two subspecies of Douglas-fir: coastal Douglas-fir that grows predominantly west of the Cascades (*P. menziesii* subsp. *menziesii*) and Rocky Mountain Douglas-fir that grows east of the Cascades and south to Mexico (*P. menziesii* subsp. *glauca*). Aside from where they grow, Rocky Mountain Douglas-fir can be identified by its reflexed bracts on smaller cones (the coastal variety's lie flat beside the scales) and the color of its foliage, which is more bluish gray. Classically, it is said that if you see a conifer—any conifer—west of the Cascades and call it a Douglas-fir, you would be right 75 percent of the time. It's that common. I highly recommend reading *Douglas Fir: The Story of the West's Most Remarkable Tree* by Stephen F. Arno and Carl E. Fielder to learn more about this wonderful tree.

TOP ROW: *Cone and leaf of subsp.* menziesii **MIDDLE ROW:** *Cone and leaf of subsp.* glauca; **BOTTOM:** *Douglas-fir forest*

CLOCKWISE FROM TOP: *Black spruce cones and leaves; Colorado blue spruce bark; characteristic spruce pegs*

SPRUCES

Picea

Spruces (family Pinaceae) are an ancient, prickly lineage of trees—most closely related to the pines (*Pinus*)—that tends to dominate cold, harsh environments throughout the northern hemisphere. Several species are native to the Northwest, but at lower latitudes, they really grow only in high mountain habitats, so our natives are comparatively rare in our cities and towns. As you go farther north, however, they begin to dominate at lower elevations too—they may be in your backyard in northwestern British Columbia or Alaska. One exception to this general rule is the Sitka spruce, which grows along the coast, not straying inland more than thirty or so miles until it gets to northern Washington, where its range starts to expand further east.

Spruces have a few traits that help set them apart as a genus right off the bat: most mature trees have relatively thin, gray, scaly bark that develops into small plates that can be flaked off, and their twigs are covered on all sides in (usually) sharply pointed, singly borne needles that leave a small woody peg when they fall away. Their cones are always pendulous when mature, often purple when young (see p. 84), and tend to grow high up in the canopy. Of these traits, the little pegs are the best clue because you can see them on any older twigs that have dropped their needles; no other genus has them.

TABLE 3. QUICK GUIDE TO SPRUCES

SPECIES	KEY TRAITS	FOLIAGE
Norway spruce *P. abies*	Needles sharp, 4-sided, green, point forward; cone longest of the spruces, up to 6 in. (15 cm) with rigid, triangular scales; tall, wide-spreading canopy with pendulous twigs	
Colorado blue spruce *P. pungens*	Needles long, stout, sharp, 4-sided, usually light bluish (also dark green), point away from twig; cone second longest of the spruces, 2.5-5 in. (6-13 cm), scales triangular, wavy, papery; tight, pyramidal crown, foliage not pendulous	
Sitka spruce *P. sitchensis*	Needles sharp, slender, slightly flattened, green above, blue below, shorter than Colorado blue's, 2-4 in. (5-10 cm); very tall, wide-spreading canopy with horizontal branching	
White spruce *P. glauca*	Needles sharp, whitish green on all sides, point forwards, 1-2.5 in. (2.5-6 cm); cones slender, scales rounded, woody; crown pyramidal with upswept twigs	
Black spruce *P. mariana*	Needles short, sharp, densely packed along twig; cone shortest of the spruces, up to 1 in. (2.5 cm), scales rounded, woody; tight, nearly columnar form with short limbs	

SPECIES	KEY TRAITS	FOLIAGE
Caucasian spruce *P. orientalis*	Needles short, stout, blunt tipped, dark green, densely pointed forward; cone up to 2.5 in. (6 cm) long, with woody, rounded scales; pyramidal crown, dense foliage	
Engelmann spruce *P. engelmannii*	Needles sharp, slender, pewter blue, point forward; twig pubescent; cone like Sitka's and Colorado blue's, but shorter, up to 1.5 in. (3.8 cm); tight, pyramidal crown with upswept twigs; rare	
Serbian spruce *P. omorika*	Needles long, flattened, two bright lines of stomatal bloom below; cone up to 2 in. (5 cm) long with rounded, woody scales; tight, nearly columnar crown with upturned twigs	

Comparison of spruce cones and foliage. Clockwise from top left: black spruce, Caucasian spruce, Colorado blue spruce, Norway spruce (far right), Engelmann spruce, Sitka spruce, white spruce

Norway Spruce
Picea abies

Native to the mountain forests of Europe, Norway spruce has been generously planted in parks and yards over the last century and is a common landscape tree throughout our region. It's not invasive, so you'll only find it where it's been planted.

Large evergreen up to 70 feet (21 meters) tall, with long, outstretching limbs that tend to swoop down heavily and turn back up at the tips; rough, pendulous, airy texture as branchlets appear to drape from limbs; dark green (not light blue). **BARK:** Gray, scaly, typical for a spruce; base develops significant buttressing. **LEAVES:** Up to 1 inch (2.5 centimeters) long, thin, sharp, 4 sided, generally pointed forward; green above, but with stomatal bloom on all 4 sides, giving a slight dark blue appearance; needles arise from all sides of twig like a bottlebrush; buds are small and papery. **CONES:** Longest of our spruces, at 4.5–6 inches (11–15 centimeters) long, greater than 1 inch (2.5 centimeters) in diameter; cylindrical with light brown, triangular scales that are rigid, not papery, when dry; hang pendulously from outer limbs.

SPECIES REMARKS: The normal species form of Norway spruce can resemble Sitka spruce in size and canopy shape, but it has a droopier appearance, and its cones are larger and woodier. This species has many cultivars, including weeping, columnar, and dwarf varieties. The large, woody cone sets this species apart from most similar species.

Colorado Blue Spruce
Picea pungens

Native to the southern Rockies but sickeningly common as an ornamental throughout our region, Colorado blue spruce can be found in yards, parks, and along streets. In the Pacific Northwest, it doesn't seed itself in, but sometimes it sure feels like it does.

Medium evergreen up to 60 feet (18 meters) tall, with narrowly conical, rigid habit and single stem; branchlets usually ascending when young, but larger branches often droop, their tips curved tightly upward. **BARK:** Typical of spruces, gray, scaly, with some brown spots where scales have fallen off; no fissures or ridges. **LEAVES:** Up to 1 inch (2.5 centimeters) long, stout, sharply pointed, square in cross section (rolls in fingers) *with stomatal bloom on all 4 sides*, giving it a bluish cast (bloom will naturally weather away, revealing a darker green on older, interior leaves); pointed outward from light brown to orangish twigs that are *glabrous*; buds are brown and papery. **CONES:** Light brown to beige, 2.5–5 inches (6–13 centimeters) long, borne in upper canopy; pendulous, with many flexible, triangular, papery scales with wavy margin.

SPECIES REMARKS: Look for the long, sharp, blue needles that point away from the stem, the papery cones, and the skinny overall habit to set this species apart. Other spruces grow much wider, with canopies that aren't as dense, or have needles that aren't as aggressive. Don't be fooled if you see a green tree that screams Colorado blue spruce in everything but the color; most varieties planted are selected for intensely blue foliage, but greener varieties are not uncommon.

Sitka Spruce
Picea sitchensis

A native species along the coast and farther inland as you go north of Washington, Sitka spruce is not infrequently planted as an ornamental in yards and parks west of the Cascades. It doesn't seed itself in outside its native habitat, though.

Large evergreen up to 100 feet (30 meters) tall in the landscape, with broadly conical habit, single leader, buttressed base; long branches with pendulous twigs; airy, somewhat messy canopy, but branches tend to be straight, not drooping; spread can be up to 30 feet (9 meters) or more on old trees. **BARK:** Typical of spruces, light gray, scaly. **LEAVES:** ¾ to 1 inch (2–2.5 centimeters) long, sharply pointed, spirally arranged around the twig, pointing in all directions; dark green on top, icy bluish below due to stomatal bloom *giving sprigs a 2-toned appearance*; 4 sided but flattened (does not roll in fingers easily); tend to point outward like Colorado blue spruce's, but needles are more slender and 2 toned (not bluish on all sides); often infected with gall at tip (see inset photo). **CONES:** Resemble Colorado blue spruce's, but shorter, 2–4 inches (5–10 centimeters) long; beige with many closely spaced papery scales; scales wavy and fringed.

SPECIES REMARKS: Sitka spruce is much larger both in height and spread than other common spruces in our region except Norway spruce. Norway differs by its droopier, much darker green foliage, and it has a much longer, woodier cone.

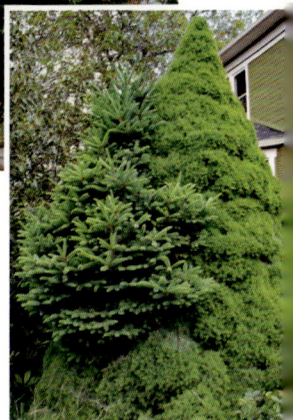

RIGHT: *Dwarf Alberta spruce cultivar with characteristic reverting shoot jutting out*

White Spruce
Picea glauca

A native of the northernmost reaches of the Northwest and beyond, white spruce is the most common species in Alaska and northern British Columbia. It's also very common elsewhere in the region but only as the lamentable dwarf Alberta spruce (*P. glauca* 'Conica').

Large evergreen up to 80 feet (24 meters) tall, with a narrow, conical habit and single stem; limbs tend to grow upward or flat, less often drooping; bluish-green overall; does not often grow into large, wide trees but stays compact. **BARK:** Typical of spruces, gray, scaly. **LEAVES:** Up to ¾ inch (2 centimeters) long, rigid, sharply pointed, square in cross section, with stomatal bloom on all 4 sides, appearing light gray; unpleasant odor when crushed. **CONES:** Light brown to beige, 1–2.5 inches (2.5–6 centimeters) long, fairly skinny even when opened; borne mostly in upper canopy, often many cones on a tree; pendulous, with many rounded (not wavy) scales.

SPECIES REMARKS: Outside of its native range, the diminutive dwarf Alberta spruce is most common—a compact, short cultivar, like a perfectly sheared Christmas tree. However, this cultivar is extremely prone to reversion, when a random shoot begins to grow as the normal species would, making it look like a real tree has sprouted right out of the canopy. It often hybridizes with Sitka spruce where their ranges overlap up north. We call this Lutz spruce (*Picea × lutzii*), and it shares traits with both parents.

Similar Species

Black spruce (*Picea mariana*) is another spruce of the far north that can sneak into some towns in its native range. You can tell the black spruce from the white spruce by its shorter needles and cones: the needles are only up to ½ inch (1.3 centimeters) long and cones up to just 1 inch (2.5 centimeters) long. It also has a very narrow, spire-like crown and grows almost exclusively in bogs.

Engelmann spruce (*Picea engelmannii*) is uncommonly planted in our landscapes but is native to high-elevation forests. If you find a tree that you think may be an Engelmann spruce planted in a lower-elevation garden or park, odds are it's really a Colorado blue spruce. But just in case, know that Engelmann spruce has *thin, sharp, pewter-blue foliage* usually less than 1 inch (2.5 centimeters) long that tends to point forward. The *newest twigs have minute hairs* on them, appearing like tiny sparkles. In contrast, Colorado blue spruce foliage is thicker and stouter and tends to point outward from the twigs, which lack hairs.

Caucasian spruce (*Picea orientalis*), another uncommon species, is tricky because its leaves aren't that sprucy. While the form, cones, bark, and needle pegs will firmly put it with the spruces, its needles are only ½ inch (1.3 centimeters) long, shiny green on top, densely packed along the twigs, and have a *blunted (not pointed) tip*; it also doesn't even *feel* like a spruce when it comes down to it. Look for its pendulous, 2.5-inch (6-centimeter), skinny, brown cones with rounded, woody scales to confirm.

Serbian spruce (*Picea omorika*) is planted more and more often in landscapes and parks due to its finer texture and tight pyramidal shape (like a shaggy column). The branches swoop down slightly, then angle up noticeably at their tips. The needles grow up to 1 inch (2.5 centimeters) long, and they're *flattened*, not angular like most other spruces; they're *green on top with two obvious bands of stomata below* (see table 3, p. 101). Cones are pendulous, generally about 2 inches (5 centimeters) long, with rounded scales.

LEFT TO RIGHT: *Subalpine fir leaves and cones; characteristic "suction cups" at the base of true fir needles*

FIRS

Abies

"Fir" is another one of those generic names for an evergreen conifer. Like "pine" and "cedar," during the colonial era, the name "fir" was slapped onto any tree that even slightly resembled the trees of Europe, even though botanically many of them are quite different. It's because of this vernacular slapdashery that we refer to trees in the genus *Abies* as "true" firs, and likewise trees in the genus *Cedrus* as "true" cedars. (It seems the pines were able to maintain their identity dominance despite the pretenders.)

Like spruces, true firs (family Pinaceae) tend to dominate forests at higher elevations, and several species are native to the Pacific Northwest—more than in any other region in North America outside Mexico (which, incidentally, has a huge diversity of conifers, the third most in the world behind the US and China, despite being only a fifth the size of either). Grand fir is the only native species that seems to prefer lower-elevation habitats where most people live, but several other native and nonnative ornamental species are planted.

True firs are characterized by regimented, upright, pyramidal forms; soft, blunt-tipped needles that attach with a suction cup–like base; and cones that grow upright (rather than hanging pendulously) and fall apart scale by scale when mature, a trait referred to as being **dehiscent**. In the Pacific Northwest, only the true cedars (*Cedrus*) have similar upright, dehiscent cones, but their needle arrangement is strikingly different. The central axis of the cones remains on the trees for some time after the scales fall away.

Grand Fir
Abies grandis

This is the most common native species in the Northwest, grown and planted throughout the region, especially at lower elevations in the western portion or in high, moist forests in the eastern portion. You'll find grand fir growing in varying habitats, essentially wherever Douglas-fir grows.

Large evergreen up to 100 feet (30 meters) tall in the landscape; can reach more than double that in habitat west of the Cascades; broadly conical through middle age with dark, dense canopy; becomes irregularly flat-topped with age, making for easy identification of old specimens. **BARK:** Dark gray, smooth when young, developing small, irregularly broken-up plates with age; thicker east of the Cascades. **LEAVES:** Up to 2.5 inches (6 centimeters) long, flattened with *notched apex* and groove down the middle on top; *distinctly 2-ranked* (especially shaded twigs) and alternating in length, with every other needle slightly shorter (like the teeth of a saw); dark, shiny green above with two stomatal lines below; divine citrus scent when crushed.

CONES: Borne upright on highest limbs, 2–4 inches (5–10 centimeters) long, cylindrical with no obvious bract (doesn't extend past scales); first purple, then beige-brown when mature before breaking apart.

SPECIES REMARKS: Grand fir's two-ranked needles (like the tines of a comb), each with two stomatal lines below, set it apart—and the scent will knock your socks off (in a good way). It can be confused with white fir (*Abies concolor*), with whom it hybridizes. Grand fir has alternating needle lengths (sawtooth) and stomatal lines below only, while white fir has uniform needle lengths and stomata above and below.

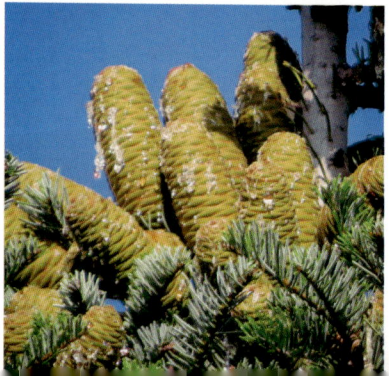

HYBRID THEORY

There are native fir trees that grow throughout the forests and towns of southern, central, and eastern Oregon that are usually called white firs. Some sources say these are true white firs and leave it at that, while others say that white fir (strictly speaking) doesn't grow much farther north than the Klamath Mountains of Northern California, and that what we call "white firs" throughout the state are in fact mostly hybrids between white fir and grand fir.

This geographic dispute in taxonomy can complicate identification. Natural hybrids can be tricky because they often display a spectrum of traits that appear in the two parent species, adding uncertainty when trying to distinguish two similar species (white and Sitka spruces in the far north of our region are another example). When trying to nail down whether you're dealing with a white, grand, or hybrid fir, consider these factors: Are you in Oregon or not? The hybrid distribution matches the state boundaries surprisingly well. If yes, where? And what do the leaves look like?

In western Oregon, Washington, northern Idaho, and southern British Columbia, grand fir is native and common, more so in urban areas west of the Cascades; its leaves are dark green on top, mostly two-ranked with alternating needle lengths. White fir is planted as an ornamental and has glaucous gray foliage that's long and curved up like a U or a V; the bark is usually thicker and rougher.

In the rest of Oregon east of the Cascades or in the Siskiyous, unless you see something that has the telltale signs of one species or the other, you're probably dealing with the hybrid (which doesn't have a formal, scientific hybrid name, although *Abies* × *grandicolor* has been floated, which is a fun suggestion). I'll call it here the fine gray fir, neither grand nor white. It'll have intermediate traits between the two parent species but likely will lean more toward white fir, with more stomatal bloom, more curved leaves, and thicker bark.

White Fir
Abies concolor

Growing natively in the mountains across much of western North America, white fir is a common component of semidry and higher-elevation forests. It's very adaptable and often found in the landscape both east and west of the Cascades.

Medium evergreen up to 60 feet (18 meters) tall, with upright, pyramidal canopy that becomes more rounded and flat-topped with age; canopy light green to grayish and somewhat layered. **BARK:** Smooth, gray when young; becomes more fissured, cracked with age, developing thick ridges separated by deep furrows, remains gray. **LEAVES:** 1.5–2.5 inches (4–6 centimeters) long, flattened, with rounded apex or very small notch; *curved upward* making a U or V shape when viewed from tip of twig; *stomata on upper and lower surfaces*, making tree appear gray. **CONES:** Similar to grand fir, but slightly longer, 3–5 inches (8–13 centimeters), with minute pubescence on scales.

SPECIES REMARKS: When identifying a white fir planted in the landscape west of the Cascades, look for long, grayish needles that are blunt tipped and curved upward; grand fir's will be green, alternating in length, and two-ranked. White fir needles are longer and less densely spaced than those of most other true firs in our region, including both native and ornamental varieties. The cone also separates it: other species have cones that are either longer or a different color, or have bracts that extend beyond the scales.

A NON-FIR AND A NON-PINE

This is a good time to mention two ornamental species from China and Japan that are victims of the poor vernacular naming of the past: China-fir and umbrella-pine. Both are found sporadically in parks and yards west of the Cascades, and both are botanically unique species that flout easy classification, causing ample confusion when you first stumble upon them.

China-fir (*Cunninghamia lanceolata*) is in the cypress family (Cupressaceae); it has orange, fibrous bark similar to other cypresses, but it has needle-like leaves that look more like the pine family (Pinaceae). Its needles are up to 2.5 inches (6 centimeters) long, flattened, and triangular with a sharply pointed tip. They are technically arranged spirally down the twig but curve and lie mostly in a messy two-ranked formation. The spiky leaves make the canopy appear prickly, like it's covered with bristle brushes. The egg-shaped cones are up to 2 inches (5 centimeters) long; they look like tight flower heads about to open, each scale a petal, but woody. Once you see one of these recognizable, unique trees, you'll know it.

Umbrella-pine (*Sciadopitys verticillata*)—an ancient conifer with its very own family, Sciadopityaceae—is a medium tree up to about 70 feet (21 meters) tall with a bushy, upright habit of dark green foliage. Its needle-like but slightly flattened leaves are light green, up to 5 inches (13 centimeters) long, and somewhat rubbery. Upon closer inspection, you'll notice the leaves appear whorled at the nodes, growing radially outward like the supports under an umbrella, hence the name. These "leaves" may be real leaves, but some botanists hold that they are instead modified stem shoots, making them technically cladodes. The true leaves, which no longer perform photosynthesis, are tiny ¼-inch (5-millimeter) scales along the main twigs. The cones are 4 inches (10 centimeters) long with thick, rounded scales that reflex backward when open.

Noble Fir
Abies procera

Another native fir that appears in the landscapes of the Northwest more often than you'd expect, as it prefers high-elevation mountains, noble fir is found as specimen trees or as a remnant of a Christmas tree farm that went feral. It's not native to lower elevations and so doesn't seed itself in.

Large evergreen up to 100 feet (30 meters) tall, with narrow, conical habit, typical of true firs; regimented, symmetrical form when young with horizontal to ascending branches that droop with age; often develops leggy, sparse canopy with dense, healthy-looking growth near the top only; bluish-green. **BARK:** Smooth, gray when young, covered with resin blisters; with age becomes more cracked, platy, usually lacks furrows. **LEAVES:** Up to 1.5 inches (4 centimeters) long, stout, blunt tipped or sometimes slightly notched, with *stomatal lines on both upper and lower surfaces*; base of needles *distinctly curved like a hockey stick* where they meet the twig on the underside; needles curve upward, concentrated above twig. **CONES:** Very large, up to 10 inches (25 centimeters) long, 3 inches (8 centimeters) in diameter, growing upright on uppermost twigs; mature to purple-gray, but covered with *long, exserted bracts* that give it a greenish-beige look; dehiscent. **SPECIES REMARKS:** Look for the cones with the pointed bracts: no other conifer has upright cones with such intense bracts. The foliage can appear similar to white fir's, but noble fir's is shorter and more densely spaced, with the telltale hockey-stick curve at the base of each needle.

Similar Species

Korean fir (*Abies koreana*) is a small tree, not more than 30 feet (9 meters) tall, often planted as a landscape tree. The flattened, somewhat paddle-shaped leaves are tightly packed along twigs and are dark green on top with bright white stomatal lines below; most cultivars planted have strongly reflexed needles that curve upside down, displaying the white stomatal lines and giving the foliage a tunneled appearance.

Nordmann fir (*Abies nordman-niana*) is another European species planted throughout the region. Its needles are about 1 inch (2.5 centimeters) long, densely arranged along the top of the twigs, and point forward as if combed. They are glossy green on top with two clear stomatal lines below and have a flat or notched tip. Cones are beige and have very slightly exserted bracts.

Spanish fir (*Abies pinsapo*) is a European species found in gardens and parks. Its short needles are unique in that they are very stout and stiff, pointing straight out from the twigs in all directions. They aren't exactly sharp like a spruce's, but they aren't soft either, with stomatal bloom on both sides, giving the tree a very pokey, metallic look. The cones are skinny, up to 6 inches (15 centimeters) long, and don't have any bracts visible (see photo in table 2, p. 87).

Subalpine fir (*Abies lasiocarpa*) and **Pacific silver fir** (*Abies amabilis*) are two confusing native species that grow in the mountains of the Pacific Northwest. They are very rarely found planted in the landscape, so be wary if you think you've found one, outside of perhaps a specialty garden or arboretum. The cone shown above is Pacific silver's, and the foliage is subalpine's whose cone can be seen on page 107.

CLOCKWISE FROM TOP LEFT: *Japanese black pine cone; characteristic pine needle sheaths; Jeffrey pine leaves*

PINES

Pinus

Pines (family Pinaceae) are some of the most ubiquitous trees in Pacific Northwest landscapes, from the wild forests across the region to the middle of the largest cities. As I've noted, a lot of trees are erroneously called pines simply because they are evergreen and have needly foliage. Given that all of the true firs, spruces, and pines were initially listed under *Pinus*, I suppose there is some precedent for the origins of such vernacular names.

Of the several native species in the region, most of them do not have a significant presence in our built landscapes. Trees like Jeffrey pine (*P. jeffreyi*) and limber pine (*P. flexilis*) are uncommon but present, while trees like whitebark pine (*P. albicaulis*) and knobcone pine (*P. attenuata*) are almost entirely absent. Only native ponderosa pine (*P. ponderosa*) and shore or lodgepole pine (*P. contorta*) could be considered common, with Asian and European species making up the bulk of the pines you'll see in cities and towns.

Pines in the Pacific Northwest are easily separable from other conifers by their needles, which are borne in bundles (called fascicles) of two, three, or five (see tables 4–6).

Each bundle has a **sheath** that wraps around all the needles at the base, like a cloth holding together a bouquet of cut flowers. When each group of needles emerges, they appear as a single needle that then splits into halves, thirds, or fifths. When they're mature, you can even realign them by sliding your fingers down the needles from their base. This characteristic belies how they can be so closely related to single-needle species like spruces and firs.

Their cones are also unique in that the exposed tip of each scale (called an **apophysis**; plural: apophyses) is quite pronounced and distinctly different from the rest of the scale, which is covered by the surrounding scales when the cone is tightly closed. Each apophysis has an accompanying **umbo**, a raised or swollen point that sometimes comes with a bonus prickle at the end.

More broadly, pines can be separated into two main subgenera: the hard pines and the soft pines. The **hard pines** are most common in Pacific Northwest landscapes. They have needles in fascicles of two or three that keep their sheaths on; their hard, woody cones have raised umbos in the *middle* of their apophyses and a "sealing band" where the scales meet while closed. The **soft pines** have needles in fascicles of five that shed their sheaths early on; their less woody, more flexible cones have umbos at the *very ends* of their scales (not in the middle of the apophysis) and lack a sealing band.

By looking for the leaf and cone traits on a given tree, you can quickly start to narrow down your options by eliminating species in the other subgenus. To further narrow down your species, take a close look at the unique appearance of the leaves, cones, bark, and form. Luckily, each species can be contrasted with the others in ways that, when done systematically, make it simple to find the right one. And remember that each cone is unique to each species, so it's perfectly possible to snag a cone and identify the tree without even looking at the leaves or form.

Comparison of two needle pine cones. Clockwise from top left: Austrian black pine, Bosnian pine, Japanese black pine, Scots pine (two cones, far right), Japanese red pine, lodgepole pine, shore pine (two closest cones at bottom), mugo pine

TABLE 4. QUICK GUIDE TO TWO-NEEDLE PINES

LEAF	BARK	CONE	
AUSTRIAN BLACK *Pinus nigra* | Common			
Long, 3–5 in. (8–13 cm), straight or curved (but not twisted), dark green; dense canopies with many dead needles	Gray, rough, with furrows, only hints of orange	2–3 in. (5–8 cm) long, green when young, with raised umbos but no prickle	
JAPANESE BLACK *Pinus thunbergii* | Uncommon			
2.5–4 in. (6–10 cm) long, twisted, dark green, with thin filaments on sheath	Gray, rough, with hints of orange	1.5–2.5 in. (4–6 cm) long, green when young, with flattened umbos	
SHORE OR LODGEPOLE *Pinus contorta* | Common			
1–2.5 in. (2.5–6 cm) long, both needles twist about 180°; shore's dark green; lodgepole's light green	Dark gray, uniformly rough along stems	1–3 in. (2.5–8 cm) long, sharp prickles on umbos, persist on tree for several years	
MUGO *Pinus mugo* | Common			
1–2 in. (2.5–5 cm) long, twisted, dark green; almost always multistemmed shrub	Dark gray, rough	1–2 in. (2.5–5 cm) long, sometimes sharp prickles, often hooked scales	

LEAF	BARK	CONE

SCOTS
Pinus sylvestris | Common

| 1–3 in. (2.5–8 cm) long, stout, twisted 180°, bluish-green; retained for approx. 6 years, thus dense canopy | Orange, flaky when young; becomes thicker, darker brown, platy with age | Up to 2.5 in. (6 cm) long, symmetrical, not retained long, with thick umbos | |

JAPANESE RED
Pinus densiflora | Semi-common

| 3–5 in. (8–13 cm) long, spreading but not twisted, bright green; concentrated near end of twigs, open canopy | Orange, flaky when young; becomes rough orange-brown with age | 1.5–2.5 in. (4–7 cm), symmetrical, persist on tree for several years, with flattened umbos | |

ITALIAN STONE
Pinus pinea | Uncommon

| Up to 6 in. (15 cm) long, muted green, pointed toward end of twig, usually dense canopy | Orange to brown; becomes platy with age | Up to 5 in. (13 cm) long, globose, with thick, rounded umbos | |

BOSNIAN
Pinus heldreichii | Rare

| 2.5–3.5 in. (6–9 cm) long, usually straight, dark green, pointed toward end of twig; dense canopy | Snakeskin-like when young; broken up into irregularly shaped plates with age | Rounded, dark blue or black when young | |

Austrian Black Pine
Pinus nigra

One of the most common ornamental species, it can be found throughout the region planted in yards, parks, and along streets. Only Scots pine can compete with it in terms of sheer numbers, but it doesn't appear to be invasive in the least.

(mugo and shore or lodgepole pines). **Japanese black pine** (*P. thunbergii*) is overall most similar, but its needles twist and have sheaths that end with two telltale, threadlike filaments, and it has smaller, more rounded cones (see table 4 and comparison on p. 115).

Medium evergreen up to 60 feet (18 meters) tall in the landscape; usually single stem with pyramidal to rounded habit, becoming flat-topped with age; dark green, often dense with dead needles in the interior. **BARK:** Develops thick plates with thin outer flakes; uniformly dark gray with only hints of orange in fissures. **LEAVES:** In bundles of 2, 3–5 inches (8–13 centimeters) long, straight or slightly curved (*not twisted*); stiff, sharply pointed, angled toward tip of twig; retained on tree for about 4 years. **CONES:** Egg shaped to conical (meaning the geometrical shape), 2–3 inches (5–8 centimeters) long; grow roughly perpendicular to twig; scales have raised umbos, but lack a prickle; *immature cones are green.*

SPECIES REMARKS: Two-needle pines similar to Austrian black pine are everywhere but differ in having orange bark (Scots and Japanese red pines) or smaller, pokier cones

INSET: *Japanese black pine needles showing sheaths that end with two threadlike filaments*

Shore Pine
Pinus contorta subsp. *contorta*

Growing natively along the coast, shore pine is the most common subspecies in landscapes west of the Cascades. You're likely to see it growing all over the place near the beach, but farther inland (but not east of the Cascades), you're likely to see it as a one-off landscape tree.

Medium evergreen, usually not much more than 30 feet (9 meters) tall, with rounded to irregular canopy, often with single, curvy stem and twisted branches; can grow as a larger tree or stay as a shrub on exposed sites. **BARK:** *Dark gray, rough,* developing small furrows and cracks between thickened, irregularly arranged plates with age; large clumps of sap common on stems in Willamette Valley due to insect damage. **LEAVES:** In bundles of 2, 1–2.5 inches (2.5–6 centimeters) long, dark green, slightly flattened, *twisted about 180 degrees.* **CONES:** Small, 1–3 inches (2.5–8 centimeters) long, asymmetric, recurved, pointing backward away from tip of twig; umbos have sharp prickle; tend to persist for several years on twig.

SPECIES REMARKS: When you see a two-needle pine along the coast that appears naturally established, it's surely a shore pine. In the landscape, though, you may confuse it with Austrian black pine or **mugo pine** (*P. mugo*), a common ornamental shrub species. However, mugo pine grows almost exclusively as a multistemmed shrub, and its cones are slightly smaller, more rounded, and often have *large, swollen, nearly hooked umbos.* Austrian black pine lacks a prickle on its cones, which are larger, and has longer leaves and lighter gray bark with large plates and furrows.

SHORE AND LODGEPOLE: ONE PINE OR MANY?

Pinus contorta is a species that really makes you believe in evolution. Absent only in the most desert-laden areas of the Southwest, it claims a wide range across western North America, from the Rockies to the Pacific and from Baja California to the Yukon and Northwest Territories. Within this expanse, it's split into about four subspecies that grow in specific areas and habitats. It's almost as if it was caught in the act of speciating, in the sense that given another few million years or so, the populations would fully diverge to the point of becoming bona fide species.

In the Pacific Northwest, *P. contorta* is represented by two native subspecies: along the coast you'll find the **shore pine** (subsp. *contorta*), a short and, well, contorted tree, and everywhere else you'll find the **lodgepole pine** (subsp. *latifolia*), which grows in the mountains, straight as an arrow. If you're near Portland, Seattle, or Vancouver, then both shore pine and lodgepole could be planted, but neither is especially common in the native forests of those areas. In my experience, shore pine is the most common in town.

LEFT: *Shore pine* RIGHT: *Lodgepole pine*

Though they are technically the same species, I've separated them into two profiles because they differ from one another in such striking ways that it's far more helpful to consider them one at a time. For a deeper dive, Michael Kauffmann goes into far more detail on all the subspecies in *Conifers of the Pacific Slope*.

Lodgepole Pine
Pinus contorta subsp. latifolia

This is the subspecies native to the mountain forests from the Cascades east throughout the rest of the region, and it's the one you're likely to find in the built landscapes there too. It tends to find its way into the landscape easily, so you may see it growing in many different contexts.

Large evergreen up to 100 feet (30 meters) tall, with upright, columnar habit and pointed top until very old age; single, straight stem with obvious whorled branching pattern. **BARK:** Brown with hints of red or brownish-orange, thinly scaled *without distinct furrows*; uniformly patterned on stems, especially as they get larger. **LEAVES:** In bundles of 2, 2–3 inches (5–8 centimeters) long, more yellowish-green than shore pine; slightly flattened and twisted. **CONES:** Small, 1–2 inches (2.5–5 centimeters) long, *asymmetric, recurved*, pointing backward away from tip of twig; scales have sharp prickle; some open when mature, most stay closed on twig for years.

SPECIES REMARKS: The most obvious differences between shore and lodgepole pine (aside from where they're growing) are the form and the bark: shore pine is a short, twisted tree with small furrows and plates, whereas lodgepole is a tall, straight tree with uniformly scaly bark and no furrows. The easiest way I have found to distinguish *P. contorta*, as an overall species, from other common two-needle pines is the bark. It doesn't have orange, flaky bark, which eliminates Scots pine and Japanese red pine; it doesn't have deep furrows and wide plates, which eliminates Austrian and Japanese black pines; and it's a tree, so mugo pine, though it has the most similar bark, is out too.

Scots Pine
Pinus sylvestris

Another ornamental from Europe that is quite common in our region, Scots pine is planted mostly in yards and "professional landscapes." It is second only to the Austrian black pine in terms of ubiquity and likely surpasses it east of the Cascades.

Medium evergreen up to 60 feet (18 meters) tall in the Pacific Northwest, with usually rounded, outstretching habit, mostly with single main stem; branch tips uniformly covered by needles; overall bluish-green foliage contrasts against orange stems. **BARK:** *Bright orange, flaky* on young stems and limbs with outer, grayer bits flaking away easily; becomes more furrowed, grayer with age as larger plates develop on oldest stems. **LEAVES:** In bundles of 2, 1–3 inches (2.5–8 centimeters) long, stiff, twisted about 180 degrees; *distinctive bluish-green* due to stomatal lines on outer sides; creates a bristle-brush appearance along twigs; needles are retained about 6 years. **CONES:** Up to 2.5 inches (6 centimeters) long, mostly symmetrical; open when mature and *not retained* like other 2-needle pines; apophyses diamond shaped; umbos often distinctly thickened, nearly hooked (but not sharp to the touch). **SPECIES REMARKS:** To set this tree apart, look for the orange, flaky bark on younger stems and branches contrasted with the bluish foliage. It's easily confused with Japanese red pine, which also has orange stems, but Japanese red pine has dozens of smaller cones that are held on the limbs for several years, as well as lighter green, less dense foliage (see p. 123 for comparison).

Similar Species

Japanese red pine (*Pinus densiflora*) looks similar to Scots pine due to its bright orange bark but is far rarer. The needles are bright green, straight, and clustered at the top of the twigs and are not held for as long as Scots pine's, plus the cone is much smaller (see table 4).

Scots pine leaf on the left and Japanese red pine leaf on the right

Italian stone pine (*Pinus pinea*), a Mediterranean species found as a landscape tree west of the Cascades, is famed for its decurrent habit, which often develops a flat-topped crown resembling a beach umbrella (see photo on p. 21). It's got muted orange bark when young that becomes browner and plated with age. Its needles are slender and long, up to 6 inches (15 centimeters), and have a muted green color. Its cones are large, up to 5 inches (13 centimeters) in diameter, globose, with thickly rounded apophyses—very distinctive for any pine in our region (see table 4).

Bosnian pine (*Pinus heldreichii*) is reportedly being planted more often as a landscape or street tree, though it's still a rare find in my experience. It looks very similar to Austrian black pine but is set apart by its diamond-plated young bark and its older bark that is uniformly broken up into small, irregularly shaped plates by shallow cracks (not deep furrows). Its leaves are shorter, and its *cones are a dark bluish-black when developing* (see table 4), whereas Austrian black pine's are green.

A few other two-needle pines are peppered into our landscapes, such as **maritime pine** (*Pinus pinaster*) and **Table Mountain pine** (*Pinus pungens*). Both are rare, mostly found in specialized collections or sometimes randomly in yards or along freeways. Look at these options if none of the others quite fit.

LEFT: *Maritime pine cone* **RIGHT:** *Table Mountain pine cone*

TABLE 5. QUICK GUIDE TO THREE-NEEDLE PINES

LEAF	CONE	
PONDEROSA *Pinus ponderosa* | Common everywhere		
Long, 8 in. (20 cm), dark green, radially spread, usually not twisted	Mostly symmetrical, smallest of the three-needle pines 2–6 in (5–15 cm), with small, sharp prickle pointed out	
JEFFREY *Pinus jeffreyi* | Not common		
Long, 5–11 in. (13–28 cm), bluish-green, usually twisted and held on upwardly pointed twigs	Large, 5–11 in. (13–28 cm) long, but not heavy, prickle not pointed out (doesn't poke your hand like ponderosa's), "Gentle Jeffrey"	
COULTER *Pinus coulteri* | Not common		
Long, 6–12 in. (15–30 cm), radially spread, bright green	Huge, over 12 in. (30 cm) long, very heavy, asymmetrical, with massive hooks pointed outward	
GHOST *Pinus sabiniana* | Not common		
Long, 8–12 in. (20–30 cm), droopy, wispy, blue-gray, canopy upright, but multistemmed	Big, heavy, but only up to 10 in. (25 cm) long, more rounded than Coulter's; hooks tend to point more downward	
MONTEREY *Pinus radiata* | Common only on southern coast		
Shortest of the three-needle pines, 3.5–6 in. (9–15 cm), green, canopy irregularly rounded	Up to 6 in. (15 cm) long, stays tightly closed most of the time, remains on tree for many years, no significant hooks or prickles	

Ponderosa Pine
Pinus ponderosa

The most common pine east of the Cascades, hands down, and easily in the top five west of the Cascades, ponderosa pine grows natively all over the region and can be found as a remnant tree in the landscape or a newly planted tree where drought tolerance is needed.

Large evergreen up to 100 feet (30 meters) tall, with pyramidal habit in youth and middle age, developing a rounded top in time; almost always with single massive stem; needles long, concentrated near ends of the twigs, giving a pillowy texture to twigs and canopy. **BARK:** Famously beige-yellow to reddish-orange when old, broken up into large, flat plates covered in small, puzzle-piece-like scales that flake off easily; younger stems more gray-brown, furrowed; trees west of the Cascades tend to keep this younger look. **LEAVES:** In bundles of 3 (rarely 2), green, up to about 8 inches (20 centimeters) long, radiating out in all directions from end of twig, sometimes drooping downward, especially in winter. **CONES:** 2–6 inches (5–15 centimeters) long, up to 3 inches (8 centimeters) in diameter, egg shaped, mostly symmetrical; do not have a stalk; are not significantly longer than they are wide; *each scale tipped with a straight prickle* that pokes you even when the cone is open.

SPECIES REMARKS: As it's the most common, if you see a three-needle pine, it's probably a ponderosa. Note that trees on both sides of the Cascades can appear slightly different, especially their bark.

Bark can appear slightly different on the east (left) and west (right) sides of the Cascades.

Similar Species

Coulter pine (*Pinus coulteri*) is famous for its massive, spiked cones, one of the biggest of all conifers, which is the easiest way to identify it (see table 5). It's planted mostly in parks and maybe some landscapes, but rarely around buildings or cars because its cones can crush things. Its needles are long, 6–12 inches (15–30 centimeters), stiff, and spread out radially from the stem but don't tend to droop intensely.

Ghost pine (*Pinus sabiniana*), also called gray or foothills pine (which are way more boring than ghost pine), is another species with a huge cone, but not quite as big as Coulter pine's (see table 5). Its needles are nearly the same length, but they are blue gray and droop distinctly, giving the tree an overall appearance of a wisp of smoke. It's almost always multistemmed, so its silhouette gives it away.

Jeffrey pine (*Pinus jeffreyi*) is native to the Siskiyous of Southern Oregon and can be found sparingly in the landscape. It looks very similar to ponderosa pine, so it takes a close eye to tell them apart (see table 5). The longer, more bluish needles and larger cone with recurved (not straight) prickles are the best clues in the landscape. The fissures in its bark smell sweetly of vanilla, while ponderosa smells of turpentine.

Monterey pine (*Pinus radiata*) is found sparingly along the southern coast of Oregon and every now and then in a city west of the Cascades. Its crown shape is a lot like shore pine's but much larger and with three needles per fascicle rather than two. Its egg-shaped cones are up to 6 inches (15 centimeters) long, intensely asymmetrical, and serotinous, meaning they stay tightly shut until intense heat, like fire, opens them. Their bark is dark gray, thick, and furrowed, creating a menacing appearance.

Clockwise from top left: Coulter pine, Jeffrey pine, Monterey pine, ghost pine

TABLE 6. QUICK GUIDE TO FIVE-NEEDLE PINES

LEAF	CONE	FOLIAGE
WESTERN WHITE *Pinus monticola* | Common		
4 in. (10 cm) long, fairly stiff, bluish-green, with 4–5 stomatal lines on inner surface	Long, 6–10 in. (15–25 cm), scales near stem recurve strongly	
EASTERN WHITE *Pinus strobus* | Common		
Similar to western white's, but softer, greener, with fewer stomatal lines on inner surface	Like western white's, but shorter, 6–8 in. (15–20 cm), scales near stem usually not strongly recurved	
HIMALAYAN WHITE *Pinus wallichiana* | Not common		
Longest of the five-needle pines, 8 in. (20 cm), noticeably droopy	Long, 6–10 in. (15–25 cm), on long stalks, with striations on scales, resinous	
JAPANESE WHITE *Pinus parviflora* | Somewhat common		
Short, up to 3 in. (8 cm), blue-green, twisted, bunched along twig	Up to 3 in. (8 cm) long, *sessile*, in clusters of usually 2–3, obvious striations on scale tips	

LEAF	CONE	FOLIAGE
HAINAN WHITE *Pinus fenzeliana* | Somewhat common		
Similar to Japanese white's	Similar to Japanese white's, but usually *solitary with ¾ in. (2 cm) peduncle*	
LIMBER *Pinus flexilis* | Somewhat common		
Up to 3.5 in. (9 cm) long, bright green to bluish, densely packed along twig, soft; flexible twigs	3–6 in. (8–15 cm) long, with reflexed cone scales	
ROCKY MOUNTAIN BRISTLECONE *Pinus aristata* | Uncommon		
1–2 in. (2.5–5 cm) long, tightly packed and closely held together; *resin dots on leaves; furry appearance*	Up to 4 in. (10 cm), a lot of tight scales with a sharp prickle on tips	

Comparison of cones for five-needle pines, clockwise from left: limber pine, western white pine, eastern white pine, bristlecone pine, Japanese white pine, Hainan white pine

Western White Pine
Pinus monticola

A native forest tree that finds its way into our built landscapes with varying frequency, western white pine is not widely common but not altogether rare either. You'll only see it as a landscape tree, as it doesn't stray from its mountain habitat without human intervention.

Large evergreen up to 100 feet (30 meters) tall, with pyramidal habit and long, horizontal branches and single stem; crown often irregularly shaped when mature, becoming flat-topped with age. **BARK:** Light gray, very smooth when young, developing distinctive, small, irregular plates that cover it uniformly. **LEAVES:** In bundles of 5, up to 4 inches (10 centimeters) long, straight, somewhat stiff (in comparison to eastern white pine), bluish-green, with *4–5 stomatal lines on lower side* (may need magnifying glass to see them); grow out radially from the stem, but aim forward near tip; like brushes (see table 6). **CONES:** Long, 6–10 inches (15–25 centimeters), with long, flat scales that bend outward so tips are nearly flat when cone is held by the small stem (peduncle); *scales right near the stem recurve strongly;* like other soft pines, umbo sits at the end of the scale, not in the middle of the apophysis; cone usually slightly curved, looking like a scaly banana when developing.

SPECIES REMARKS: Most easily confused with **eastern white pine** (*P. strobus*), western white has longer cones with recurved scales near the peduncles and 4–5 stomatal lines on the inside of each leaf compared to eastern's two. Eastern white pine's foliage is overall much thinner, softer to the touch, and bunched near branch tips. **Sugar pine** (*P. lambertiana*) is rarely found outside the forest but looks very similar to western white pine except its cones are far larger and woodier (see cone on p. 130) and its leaves are uniformly covered in stomatal bloom.

Western white pine cone (left) and eastern (right)

129

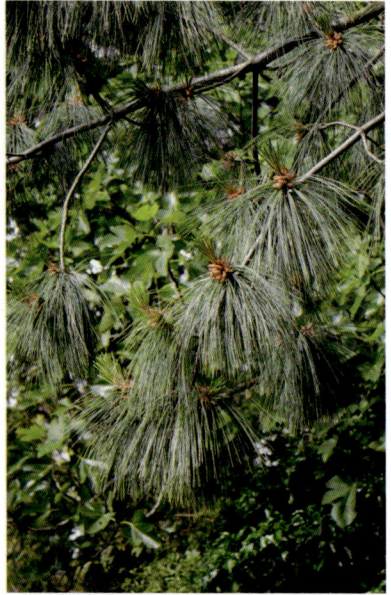

Similar Species

Himalayan white pine (*Pinus wallichiana*) is much rarer in our landscape, but you'll inevitably come across it and wonder what it is. The most striking features are its needles and cones, both of which are much longer than those of the other more common five-needle pines. The needles, which will surely set it apart, are up to 8 inches (20 centimeters) long and very droopy, giving the tree the appearance of a blurry photo. Its cones are similar to western white pine's, but wider, with larger scales that have more pronounced striations on the tips, and a longer stem.

TOP: *Himalayan white pine cone and foliage*
RIGHT: *Sugar pine cone*

Japanese White Pine
Pinus parviflora

This tree, which has become popular recently as a small, striking landscape tree, is native to Japan and Korea and well-suited to our climate west of the Cascades. It's planted often in yards and landscapes where a tree that isn't going to outgrow its quarters is wanted.

Medium evergreen up to 50 feet (15 meters) with pyramidal, but layered and bunchy, habit; branches tend to be well separated from one another; foliage is tightly packed in pom-pom-like tufts, giving the tree a sculpted look. **BARK:** Brownish-gray, smooth on young trees, becoming rougher, slightly furrowed with age; often has peeling scales over larger sections broken up by irregular furrows. **LEAVES:** In bundles of 5, short, only up to 3 inches (8 centimeters) long; stout, twisted, held closely together in their bundles; whitish-blue due to stomatal lines on underside exposed by the twist in needles. **CONES:** *Sessile*, up to 3 inches (8 centimeters) long, *in clusters of 2–3 all over newer shoots*; scales wide, rounded, with obvious striations on outer side; look like other white pine cones, only miniaturized.

SPECIES REMARKS: The short, stout, bluish-green needles that give the whole tree a distinctive look are the best way to separate this species from others. We almost exclusively plant the shorter-leafed cultivar 'Glauca'. The very similar, closely related **Hainan white pine** (also called **Kwangtung pine**, *P. fenzeliana*; syn. *P. kwangtungensis*) is easily confused with this species, but its cones are *solitary and have a ¾-inch (2-centimeter) long peduncle.*

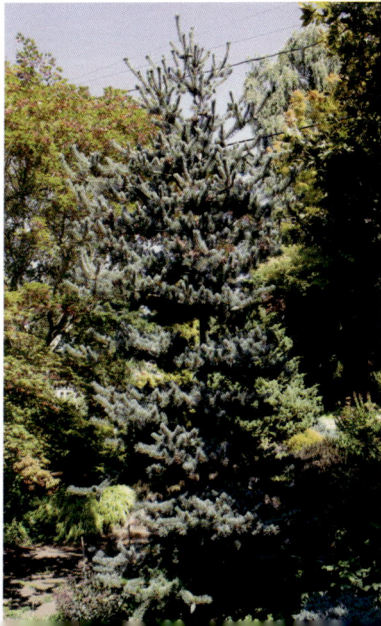

BELOW: *Hainan white pine cone: smaller, solitary, with a long peduncle*

131

Limber Pine
Pinus flexilis

Though native in the mountains through-out the west, like western white pine, limber pine doesn't stray into the built landscape by itself, but rather is planted more often as a tough, drought-tolerant species for streets and landscapes across the region.

Medium evergreen up to 50 feet (15 meters) tall in the landscape, but usually smaller; single stem and columnar habit when young; tends to spread out and develop a more decurrent growth form with age. **BARK:** Smooth, gray when young, becoming very rough, flaky with age; dark gray outer bark with lighter orange under-neath and where flakes have fallen away. **LEAVES:** In bundles of 5, up to 3.5 inches (9 centimeters) long, bright green, fairly dense along twigs (compared to young western and eastern white pines, which are sparser); needles soft; twigs distinctly flex-ible (hence name), able to literally be tied in a knot without breaking. **CONES:** 3–6 inches (8–15 centimeters) long, skinny, with thickened, reflexed, woody scales that look like the spout of a pitcher.

SPECIES REMARKS: 'Vanderwolf's Pyr-amid' is the most common cultivar; it has bluish foliage and a very upright habit that becomes rounded with age. It looks similar to Japanese white pine but has a denser can-opy, longer needles, and longer cones with reflexed scales. Its stomatal lines on both sides of the needles are somewhat indistinct, which further separates it from eastern and western white pines, which have pronounced stomatal lines only on one side.

Rocky Mountain Bristlecone Pine
Pinus aristata

This unique species is found sparingly in the southern and western towns of the Pacific Northwest but more often as you go north and east, as it can take (or rather prefers, it seems) cooler and/or drier conditions. It is planted mostly as a landscape tree or rarely as a unique street tree.

Small evergreen up to 30 feet (9 meters) tall, most often with very irregular canopy, looking like a series of snaking, furry limbs growing in all directions. **BARK:** Smooth when young, but most often concealed by foliage; becomes a rough gray-brown with age, developing slightly flaky scales. **LEAVES:** In bundles of 5 but sometimes appearing as just one thick needle because they stay tightly together, especially when first emerging; 1–2 inches (2.5–5 centimeters) long, stiff, aimed toward tip of twig;

dark green outside with distinctive resin dots unlike any other conifer's in our region; needles held for up to 15 years, making branches look densely woolly. **CONES:** Up to 4 inches (10 centimeters) long, conical, symmetrical with many closely spaced scales, each tipped with a long, *bristly prickle at end of umbo.*

SPECIES REMARKS: The relatively short needles persist for many years along the branches, which gives them a snaky, fuzzy appearance. That and the unique resin dots on the leaves set it apart. No other bristlecone pines are found in our region.

133

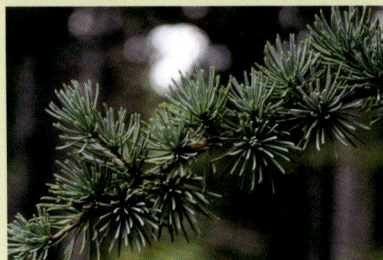

CLOCKWISE FROM TOP LEFT: *Cedar of Lebanon cones; comparison of cedar leaves: Atlas (top); Lebanon (center); deodar (bottom); cedar of Lebanon branch*

CEDARS

Cedrus

I believe the cedars (family Pinaceae) cause the most confusion when people first learn to identify trees using their scientific names because "true" cedars don't look anything like what most people commonly call cedars—namely, those scale-leaf trees in the cypress family. "Cedars" exemplify the dissonance between common and scientific names and thrust you headfirst into the deep end of the world of scientific taxonomy.

This shouldn't be seen as some kind of shortcoming, by the way. Like Neo entering the Matrix, when you're told that everything you think you know about a "cedar" is false—literally—and that a tree that you're pretty sure is a pine is in fact a *true* cedar, and there's a whole secret system that defines this that you're only starting to learn, it can be understandably jarring.

The genus *Cedrus* was given to a group of trees with a specific botanical description, and anything that strayed too far from those particular traits was definitionally not a *true* cedar. True cedars, of which there are three accepted species, have large cones that grow upright on short spur shoots, falling apart scale by scale when mature.

True cedars also have needle-like foliage, not scale-like. Their needles are singly borne on new shoots but then grow from short spur shoots after the first year, appearing as if they are borne in clusters of fifteen to forty. This is just a trick, however, caused by strong dimorphic growth. The primary shoots (elongating twigs) grow long and quickly, setting single needles in an easily discerned spiral pattern along the stem as they go. Secondary shoots (spur shoots) do the same but grow very slowly, such that the spirally arranged needles are so close to one another that they appear to be growing from the same node. So yeah, they can be pretty confusing at first.

Deodar Cedar
Cedrus deodara

It's tough to say which of the true cedars is the most common, but my money is on this one. Also known as Himalayan cedar, it's planted across the Northwest landscape but is most prevalent on the west side of the Cascades, from business parks to yards and streets to old parks.

Large evergreen growing to over 70 feet (21 meters) tall, but often quite variable in height and spread; pyramidal when young; tends to develop reiterations from branches that turn upward and form new leaders with age; *branches tend to be horizontal but drooping near tips*. **BARK:** Gray, rough, quickly breaking into flat-topped plates, like ice sheets breaking apart on the ocean; becomes rougher, more deeply fissured with age. **LEAVES:** About 1.5 inches (4 centimeters) long, singly borne, spirally arranged on new shoots and lying down along stem near tip, pointed more outward as you move up the twig; most foliage borne on short spur shoots, densely clustered together with 15–20 per cluster; generally soft, pliable, dark green with bluish tinge and waxy bloom, giving foliage a light silvery color. **CONES:** Solitary, barrel shaped, growing upright on short shoots, 3–4 inches (8–10 centimeters) tall; scales appear as tightly packed layers and remain that way until cones mature, dry, and fall apart scale by scale.

SPECIES REMARKS: The unique form makes for a characteristic tiered appearance, and the upright, dehiscent cones paired with the clusters of needles set this species apart. Other cedars' needles are shorter and stiffer.

Atlas Cedar
Cedrus atlantica

Though deodar may be more common overall, Atlas cedar is certainly exceeding it in new plantings due to its attractive foliage and range of cultivars. It's planted in new landscapes in its tree form, but perhaps just as often as a snaking, pendulous form or skinny aberration of a tree.

Large evergreen growing to over 60 feet (18 meters) tall in the landscape; most cultivars very upright with strong central leader, pyramidal habit, ascending branches; texture very sharp, like a tree covered in pincushions. **BARK:** Similar to other cedars; gray, rough, breaking into flat-topped plates; very uniformly covering stem and larger branches. **LEAVES:** Similarly arranged on new shoots and short shoots as on deodar, but shorter, ¾ to 1 inch (2–2.5 centimeters) long, and stiffer; generally more needles per cluster (30–40); greenish-blue to bright whitish-blue, covered in glaucous waxy covering. **CONES:** Solitary, barrel shaped, growing upright on short shoots; up to 3 inches (8 centimeters) tall, but otherwise very similar to other cedars.

SPECIES REMARKS: Atlas cedar also has a unique form, but it's basically opposite that of deodar cedar: instead of having horizontal limbs with droopy twigs, Atlas's twigs are rigid and held on upright branches. The cultivar 'Glauca' is most commonly planted; its icy blue foliage is very distinct, making it quite unmistakable from deodar cedar.

WAIT, HOW MANY CEDARS ARE THERE?

I was afraid you'd ask. Today there are three generally accepted species, the two just described in detail and **cedar of Lebanon** (*Cedrus libani*). However, depending on your source (namely, if they're a lumper or a splitter), there are as few as two and as many as four. In the former case, Atlas cedar is lumped in as a subspecies of cedar of Lebanon (*Cedrus libani* subsp. *atlantica*), along with another subspecies, the not-at-all-confusingly named cypress cedar (*Cedrus libani* subsp. *brevifolia*). In the latter case, each is split into its own species. Most taxonomists agree that there's enough evidence to keep Atlas cedar as its own species but cypress cedar as a subspecies.

Cedar of Lebanon is planted in the Pacific Northwest and appears for all intents and purposes as a darker green version of Atlas cedar, with branches that are neither too upright nor too droopy but grow more or less horizontally. Closer inspection will show you that its twigs grow more densely together and are usually hairless (or nearly so), while Atlas has more downy shoots and a less dense canopy overall. A fun way to remember which form is which is to use this mnemonic "*A* is for ascending Atlas, *L* is for level Lebanon, and *D* is for droopy deodar."

Now, if you're inclined to be a splitter, and believe in the cypress cedar as its own species (promoted to *Cedrus brevifolia*), then you may find some planted in our region too. It's distinguished by its height, not growing much more than 40 feet (12 meters) tall, and by its very short needles, which don't get much more than ½ inch (13 millimeters) long. If you're a lumper, though, and you don't believe in it, then you're only going to find a diminutive version of cedar of Lebanon with short leaves. And that, my friends, is the joy of taxonomy!

CLOCKWISE FROM TOP: *Japanese larch in fall; comparison of larch cones, clockwise from center top: Japanese larch, western larch, Siberian larch (two cones), American larch (single cone, bottom left), and European larch (three cones, left side); European larch leaves*

LARCHES

Larix

Larches (family Pinaceae) are another unique group in the Northwest because, like the dawn redwood and baldcypress, they are deciduous conifers, shedding their needles fully in the late fall. In wild forests, they put on quite a show in October; our two native species turn bright orange-yellow among a backdrop of dark evergreen conifers in the mountains.

Larches produce a very telling texture due to their dimorphic twig growth. Like the cedars, larches produce needles clustered on short, peg-like spur shoots. When they lose their leaves in the fall, scraggly, bumpy twigs crisscross and interweave, forming their winter silhouettes, unique to larches. They also have upright-growing cones, but the cones are not dehiscent and look quite different from cedar cones.

Western larch is found sparingly in the landscape even though it grows natively in the mountains from the Cascades east to the Rockies of western Montana. (The other native species, alpine larch, *Larix lyallii*, doesn't make it out of the highest, most extreme mountains.) The rest of the larches are represented by a handful of ornamental species from Europe and Asia, as well as the American larch, which can only be claimed as native to the boreal forests of northern Canada and Interior Alaska (see table 7).

TABLE 7. QUICK GUIDE TO LARCHES

LEAF	CONE	
WESTERN		
Larix occidentalis | Somewhat common		
15–30 per spur, ¾ to 2 in. (2–5 cm) long, light green; pubescent twigs at first	Up to 2 in. (5 cm) long; *bracts extend beyond scales*	
EUROPEAN		
Larix decidua | Common		
30–40 per spur, 1–1.5 in. (2.5–4 cm) long, shorter than Japanese's, dark green; twigs mostly hairless, *straw yellow* in winter	1–1.5 in. (2.5–4 cm) long, no bracts visible, scales *not intensely recurved*	
JAPANESE		
Larix kaempferi | Common		
40 or so per spur, usually slightly longer than European's, green; *twigs purplish-pink* in winter	Similar to European larch's, but *ends of cone scales recurve back*, giving rosette appearance	
AMERICAN		
Larix laricina | Somewhat common		
12–30 per spur, shorter needles, only up to 1 in. (2.5 cm) long, bluish-green; twigs orange-brown, hairless	Smallest of the larches, like rosebuds; not more than 1 in. (2.5 cm) long, fewer scales per cone than others	
SIBERIAN		
Larix sibirica | Common in north		
1–2 in. (2.5–5 cm) or longer, green; *twigs pubescent*	Rotund, wide and rounded cone scales that are pubescent with striations	

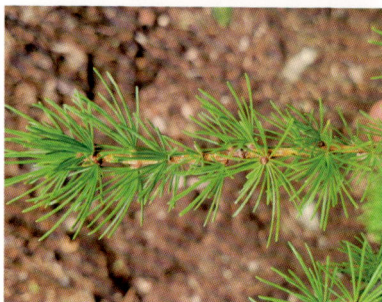

Western Larch
Larix occidentalis

It may not be the most common larch in the landscape, but the western larch is our native species and quite different from the others, so it gets to go first. You'll see it planted in the landscape as a specimen tree in yards and parks, and every now and then as a street tree, but almost nowhere else.

Large deciduous conifer, growing to over 165 feet (50 meters) tall in habitat, but much smaller in the landscape; single stemmed with pyramidal habit and ascending, spiky, somewhat unkempt branches; pronounced light green color; sparse, bumpy twigs in winter. **BARK:** Thick, furrowed, similar to Douglas-fir on older trees but with more orange tones. **LEAVES:** ¾ to 2 inches (2–5 centimeters) long, spirally arranged on new shoots; after first year, borne in clusters of *15–30 needles* on short pegs; needles light green, soft, pliable, lacking waxy coating of evergreen needles; crown often appears airy; twigs orange-brown, pubescent when first emerging, becoming less so with age. **CONES:** Up to 2 inches (5 centimeters) long, solitary but numerous, borne more or less upright on short shoots and usually retained along length of stem for several years after opening; have *obvious, skinny bracts extending beyond cone scales* (which can often fall away after first year, causing confusion, so look for multiple examples and newer cones).

SPECIES REMARKS: To set this species apart, look for the extended bracts on the cones; all other larches have bracts that are shorter than the scales when the cones are mature, so they appear absent. The form can also be helpful, as western larch grows horizontal or ascending twigs, while the others have more drooping branchlets.

European Larch
Larix decidua

This species is common in parks and landscapes as a tree of interest. Like with most larches, its form in the landscape appears a bit messy, but this one tends to stay the cleanest. It's about as common as Japanese and western larches and doesn't seed itself in.

Medium deciduous conifer up to about 60 feet (18 meters) tall in the landscape; single stemmed with pyramidal habit at first, tending to become more irregular with age, yet still retaining single dominant stem; *branchlets tend to droop.* **BARK:** Between furrowed and flaky, gray-brown, with orange tones where outer bark has flaked off or in furrows. **LEAVES:** 1–1.5 inches (2.5–4 centimeters) long, spirally arranged on new shoots; after first year, borne in clusters of *30–40*

needles on short pegs; needles soft, pliable, darker green than western larch; twigs pale straw color in winter, mostly hairless. **CONES:** 1–1.5 inches (2.5–4 centimeters) long, solitary but numerous; borne upright on short, curved shoots; retained for several years after opening; *lack an exserted bract beyond cone scales; scales open to be mostly flat, but do not reflex backward intensely (see table 7).*

SPECIES REMARKS: Larches are hard to tell apart; it takes observing a few unique parts and some patience to nail them down. The straw-colored twigs and non-reflexed cone scales set European larch apart from Japanese larch, which has purplish-pink twigs and scales that reflex dramatically. Its bractless cones set it apart from western larch, and its cone size and number of scales separate it from American larch.

Japanese Larch
Larix kaempferi

The more I look, the more I see this species in the landscape, especially as a yard and park tree. It's still not especially common, but certainly west of the Cascades it's a popular ornamental. It doesn't seed itself in, so you'll only see it where planted.

Large deciduous conifer up to 90 feet (27 meters) tall, with pyramidal habit that becomes wider and more open with age; crown doesn't get too dense; drooping branchlets hang off large, horizontal to slightly ascending limbs. **BARK:** Similar to European larch; gray with flaky furrows and orange under exfoliated flakes. **LEAVES:** 1–1.5 inches (2.5–4 centimeters) long, spirally arranged on new shoots, borne in clusters of up to 40 needles on short pegs after first year; needles green, soft, pliable, often just slightly longer on average than European larch; *twigs are purplish-pink in winter*, contrasting with European larch's straw yellow. **CONES:** Very similar to European larch, but the tip of each scale reflexes backward dramatically when mature and dry, giving a rosette appearance.

SPECIES REMARKS: The twig color and reflexed cone scales (which lack bracts) set this species apart from other similar species. The form of this tree is often variable also, as I've seen several specimens that look like contorted, ancient gargoyles on one side of town and statuesque, pillar-like forest sentries on the other. Such is the variety of life.

American Larch

Larix laricina

Also called tamarack, American larch is mostly native to the boreal forests of Interior Alaska and northern Canada in our region, and it has a range that stretches farther south the farther east you go, extending to the Great Lakes area and New England. It's sparingly planted in the landscape but gets more common as you go north.

Large deciduous conifer up to 65 feet (20 meters) tall, with pyramidal habit, especially when young; single stemmed; fluffy appearance, often maintaining uniformly conical habit with age. **BARK:** Rough when young, developing into many small, flaky scales; gray revealing brownish-orange below. **LEAVES:** About 1 inch (2.5 centimeters) long, spirally arranged on new shoots, borne in clusters of *12–30 needles* on short pegs after first year; needles slightly bluish-green, soft, pliable; twigs are orange-brown, glabrous. **CONES:** Very small, usually less than 1 inch (2.5 centimeters) long, with far fewer scales (10–20) than other species; lack bracts and do not have reflexed scales; like small rosebuds along twigs.

SPECIES REMARKS: Differentiate this species by the smaller cones, shorter needles, and scalier, flakier bark. Counting the cone scales and noticing how spread out they are will help you identify it even if the cones are slightly larger than expected: cones of other species will have twenty to forty scales that spread and lie flatter, while American larch scales tend to stay more compactly arranged. **Siberian larch** (*L. sibirica*) is often planted in the northern portions of the Pacific Northwest where you'd also likely find American larch. It differs by having larger cones with bigger scales that have striations along them, and both its twigs and cone scales are pubescent when young.

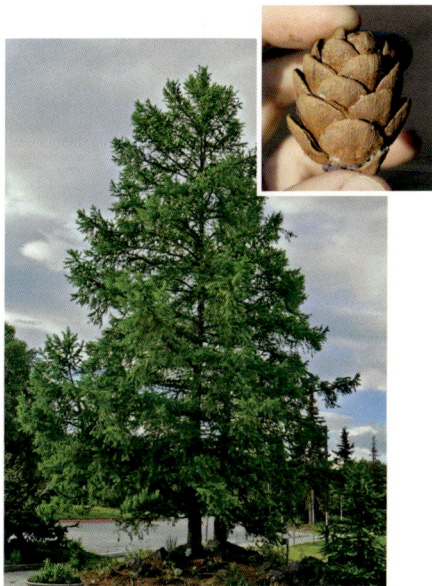

ABOVE: *Siberian larch form and cone*

143

BROADLEAF TREES

Though the natural forests of the Pacific Northwest are famous for their diversity of conifers, there are far more species of broadleaf trees in our region's cities and towns. In fact, our broadleaf trees outnumber our conifers nearly three to one in terms of species, so the real diversity in our region is found right outside our front

Broadleaf is a general, nontechnical term that refers to a biological group of plants called **angiosperms**, or more simply, plants with flowers (conifers and their kin are called **gymnosperms** as they lack flowers). Broadleaf trees, as their name suggests, tend to have wide, flattened leaves and produce various kinds of fruit with seeds squirreled away inside. They can be more challenging to tell apart than the conifers because, as I noted in the conifer introduction, you must deal not just with a single but with a flower *and* a fruit.

Genera like the oaks, lindens, maples, catalpas, and elms each have fruit that is consistent and specific to their genus: each species has similar fruit, but with slight variations on the original design. This is a great place to start if you find a telltale fruit type on a tree, as it immediately eliminates all other genera. However, some fruit is just simply not as clear-cut.

Fruit types like **drupes** or **capsules** are often simply generic descriptions, offering little help in narrowing down a particular genus or species. On top of this, these terms denote specific botanical traits that can be hard for the casual observer to parse out in the field. For example, the difference between a drupe and a **berry** as it relates to Oregon-myrtle comes down to whether the outer layer of the seed or the inner layer of the ovary is the hard, nutty shell part beneath the fleshy outer layer. Figuring out this distinction is not a useful path to go down for our purposes. However, when you do find fruit, it is helpful to pay attention to its easily deduced traits, like shape, color, and texture (e.g., if it's fleshy or dry). In these cases, the fruit is often best used as confirmation, balanced with multiple other traits.

It's my opinion that flowers are like fruit in this way, though I know that more

intense botanical types will certainly disagree. (I hereby apologize for offending you if you're one of those intense botanical types—thanks for buying this book!) Flowers are most helpful in botanizing, but for identification purposes in this guide, they are merely another trait to consider. What's most important in this context is where the flowers are growing, what time of year they grow, and how they are grouped. For example, among the cherries and dogwoods, the timing of their flowering and their grouping can be a primary means of separating species. So, focus on the macro-level characteristics first when it comes to flowers, and balance them with other traits to nail down the identity.

This section is broken down first into oppositely arranged leaves, followed by alternately arranged leaves. Those are then subdivided further based on whether the leaves are simple or compound. Beyond that, different species and genera are grouped by how similar their leaves look. One important note: if a genus is introduced, I cover all the species in that genus before moving on to another; often (but not always), closely related genera are grouped together because their leaves look similar.

I have not included a key for the broadleaf trees. There are simply too many to cover with too many overlapping traits, so any attempts would be messy and ultimately get out of hand. In lieu of a key, see the unique characteristics lists at the end of The Stepwise Journey section to help you narrow down your tree based on specific traits you may find aside from the leaves. If you find a tree with very interesting bark, or with a generic round, black fruit, check these lists for suggestions. They're meant to offer you a good starting place, but treat them as a jumping-off point.

OPPOSITE, SIMPLE

The trees in this section all have buds and leaves that are oppositely arranged. As noted in The Stepwise Journey, leaf and bud arrangement is one of the first and most important traits to look for when identifying a tree because it's very consistent within a species, and usually within a genus and family too—though the inevitable exceptions can be found, like the giant dogwood, for example (see p. 152).

I've started with oppositely arranged trees because this group is smaller compared to the alternately arranged trees, and therefore a bit easier to wrap your head around. The trees that have simple leaves and more or less pinnate veins are listed first, followed by those with palmate veins.

For this section, however, keep two things in mind. First, I've included the trees with subopposite arrangements (like katsura, crepemyrtles, and fringe-trees) and the giant dogwood (which has alternately arranged leaves), as it's where they fit best. Second, you'll find that maples are split between the section on trees with simple leaves and the section on pinnately compound leaves, as two members of that genus have compound leaves. Luckily, they end one section and begin the next, so they are still ordered together across the sections.

Historically, a fun mnemonic used in tree identification classes to help students remember which trees have oppositely arranged leaves was MADHorse, which stands for maple, ash, dogwood, and horsechestnut. Unfortunately, it's too simplified for the diversity of trees in the Pacific Northwest, as it omits catalpas, olives, fringetrees, and tree lilacs. But you can still come up with a new one that helps you remember—it's worth a try!

LEFT: *Japanese fringetree in full bloom* **RIGHT:** *The oppositely arranged leaves and buds of vine maple*

LEFT TO RIGHT: *Pacific dogwood flower; giant dogwood leaves*

DOGWOODS

Cornus

Dogwoods (family Cornaceae) are found in great numbers in the Pacific Northwest, for better or worse. In midspring, they're one of the most beautiful trees in the region, hands down. But like cherries, they are planted in landscapes, yards, and along streets as a small, pretty, flowering tree, elevated to the specimen-tree level when perhaps a larger, tougher, more substantive tree would be more suitable.

Alas, regardless of my feelings, they're everywhere, especially west of the Cascades. There are three species that most people see and immediately think "dogwood" and two others that often come as a surprise because they lack the signature dogwood flowers—or "flowers," because the pink and white displays in spring (like the one shown above) are actually created by modified leaves called bracts. The true flowers are quite small and clustered together in the center of the flower heads.

All dogwoods have leaves with acuminate veins, which curve toward the tip of the leaf, often never touching the edge at all. Only viburnums share this trait (along with oppositely arranged leaves), but they are almost never treelike, and those that are tend to have lobed leaves whereas dogwoods do not. All but one species of dogwoods in the region have oppositely arranged leaves. This single exception still has the vein pattern (see giant dogwood leaves above), so it shouldn't cause too much trouble.

TABLE 8. QUICK GUIDE TO DOGWOODS

SPECIES & KEY TRAITS	FRUIT	TWIG
Eastern flowering (*C. florida*) Leaves dull green, often folded along midrib; twigs purplish-red, undulate with onion-like buds; bark gray, rough; flowers with 4 large, notched bracts, usually pink		
Pacific (*C. nuttallii*) Leaves like eastern flowering's, but larger, more papery; twigs straight with quarter-sized, naked flower buds; bark gray, smooth; flowers large, usually 6 oval bracts, white; blooms early, along with eastern flowering		
Kousa (*C. kousa*) Leaves shiny green with very acuminate tip; twigs gray-brown with buds that look like the Taj Mahal; bark light gray, exfoliating in irregular patches; flowers with 4 slender, pointed bracts, white or pink, on long pedicels; blooms latest		
Cornelian-cherry (*C. mas*) Leaves smallest of common species, curved downwards and folded along midrib; twigs tan, straight, with small, paired buds on short stems; bark tan, rough, somewhat scaly; flowers small, yellow, no bracts; blooms earliest (late winter)		
Giant (*C. controversa*) *Leaves alternate, oval*, widest of dogwoods, dark green but not shiny; undulate twigs with a single red terminal bud; bark gray, smooth to cracked; flowers white, in large clusters (umbels), no bracts		

Eastern Flowering Dogwood
Cornus florida

This is the most common dogwood overall, planted all over the landscape but mostly in yards and along streets. It's an old-fashioned tree, popular during the housing booms of the last century, so often large, old trees are found near old houses.

Small tree up to 25 feet (8 meters) tall, slow growing with rounded, dome-like canopy; doesn't maintain central leader for long, quickly splitting into several main scaffold limbs. **BARK:** Dark gray, rough, broken into angular sections; resembles alligator skin. **TWIGS:** Slender, purplish-red with glaucous bloom; *often appear to undulate*, slightly swooping down and then up again; buds of 2 kinds: slender, pointed vegetative buds and round, bulbous flower buds that *look like tiny terminal onions*; these unfold into bracts that swell to become the showy parts in spring. **LEAVES:** Usually up to 4 inches (10 centimeters) long, oval, with *pronounced, acuminate veins*; margin entire, apex pointed; often curved downward. **FLOWERS:** Tiny, almost inconspicuous, clustered together in center of bloom, surrounded by 4 large, showy bracts, *each with distinctive notch at apex*, usually pink in the Northwest, but sometimes white (the split-color tree below is not normal, but it's pretty cool); first to bloom. **FRUIT:** Small red drupe, 1 produced per individual (actual) flower; 2–4 clustered in center of spent flower head in fall.

SPECIES REMARKS: Regardless of the bract color, this species always has four bracts with a notch at their tips. Its rough, angular-patterned bark also sets it apart from all other species.

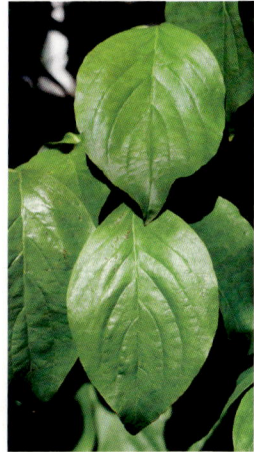

Pacific Dogwood
Cornus nuttallii

Our amazing native dogwood is not as common as its eastern cousin, but neither is it rare. It grows natively in forests west of the Cascades, so you're just as likely to see it growing in the landscape as you are to see it peeking its head out from roadside forest edges.

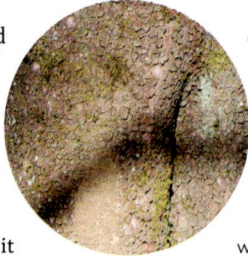

Medium tree up to 40 feet (12 meters) tall, with upright habit and dominant central leader (not rounded like eastern flowering dogwood). **BARK:** Smooth, gray when young, developing cracks and small plates with age; does not exfoliate or become intensely rough. **TWIGS:** Like eastern flowering dogwood but not as densely arranged; tipped with large *naked flower buds* about 1 inch (2.5 centimeters) in diameter, with tiny bracts around edges that will swell to become the showy parts in spring. **LEAVES:** Up to 5 inches (13 centimeters) long, oval, with characteristic acuminate veins and generally wavy margins; more rounded, less pointed than others. **FLOWERS:** Predominantly white, with 6 (sometimes 4–8) *unnotched* bracts; bracts are larger, rounded, often overlapping with one another, creating a very full bloom; bloom early in spring. **FRUIT:** Tiny red drupes that swell together in the fall.

SPECIES REMARKS: The smooth bark and upright form should set this species apart quickly. In summer, look for the larger, floppier leaves, and in winter, the exposed flower bud: other species' buds will be much smaller and wrapped in protective bracts.

Kousa Dogwood

Cornus kousa

The second most common dogwood in the landscape, kousa is native to China, Japan, and Korea. It is planted wildly as a tough, quicker-growing alternative to the eastern flowering dogwood. (It's also resistant to dogwood anthracnose, which heavily affects eastern flowering and Pacific dogwoods.)

Small tree up to 25 feet (8 meters) tall, with rounded canopy; tends to be more irregularly rounded than eastern flowering, with more horizonal branches. **BARK:** Gray, exfoliating away, leaving camouflage pattern of grays, browns, and oranges. **TWIGS:** Slender and gray-brown with hints of red, generally straight, glabrous, but with some bloom; appear armored (to me); buds of 2 kinds: slender, pointed, valvate vegetative buds and rounded, more bulbous flower buds that are pointed at the tip. **LEAVES:** Up to 4 inches (10 centimeters) long, ovate, with *pronounced, acuminate veins*; margin entire, slightly undulate, with long acuminate tip; shinier, waxier than others. **FLOWERS:** Last to appear in late spring, nearly 3 weeks after eastern and Pacific dogwoods; sit up on long pedicel and have *4 bracts*, usually white with some pink, very pointed. **FRUIT:** Compound drupe, a red bumpy sphere in fall on a long stalk; edible and distinctly different from others.

SPECIES REMARKS: The flowers and leaves are more slender and much pointier than those of the other common species, and the fruit and bark are unique to it alone. Compare the buds, bark, and fruit to ensure this is the species.

Similar Species

Cornelian-cherry (*Cornus mas*) will not strike you at first as a dogwood because its small yellow flowers lack any floral bracts; when the leaves come out, however, the similarities are obvious. It's often one of the very first trees to bloom in late winter, producing bursts of yellow from paired buds in the axils of the twigs. The

HYBRID DOGWOODS

Eastern flowering, Pacific, and kousa dogwoods can all be easily differentiated from each other by their bark, flower, and bud traits. I used to include bloom time with this list, but I have since found that the numerous hybrids have thrown that useful characteristic out the window. You can find many cultivars and hybrids between these three species that have different bloom times, often a mix of the two parent species' times (especially if kousa is one parent, as it blooms several weeks after the other two).

If you're using bloom time as a tell for which species it is, then depend on the extremes: eastern flowering dogwood and Pacific dogwood bloom first, and kousa blooms last. Anything in between is likely some kind of hybrid, and it'll probably have some mixture of traits between the parents. *The Tree Book* by Michael Dirr and Keith Warren is a helpful guide for picking these hybrids apart.

buds remind me of the eyes of a crab, little spheres on stalks, opening slightly in late winter with a sneak peek of the flowers to come. Its leaves look like those of eastern flowering dogwood but are smaller and more folded. It's a small tree, often no taller than 20 feet (6 meters) and with a rounded canopy, usually planted as a street tree but rarely elsewhere and never in natural areas. It produces singular, red, oblong drupes that look superficially like cherries, though they ripen in the fall.

Giant dogwood (*Cornus controversa*) is the only common treelike dogwood species in our region that doesn't have oppositely arranged leaves (see photo p. 147). Controversial indeed! (Pagoda dogwood, *C. alternifolia*, also has alter-nately arranged leaves, but it's rarer and stays shrubby most of the time.) It gets much larger than most dogwoods—up to 50 feet (15 meters) tall—and has larger, more heart-shaped leaves. Its branches grow horizontally and undulate as they spread out, creating a unique layered appearance. The flowers also lack bracts, but instead of being small and clustered together, they are spread out in a large, flat-topped corymb resembling a viburnum. They appear in late spring and produce small berrylike drupes in the fall that change from red to purple to nearly black. Look for the empty fruit clusters still hanging on at the ends of twigs, as well as the undulating, horizontal pattern of growth in the winter.

LEFT & MIDDLE: *Cornelian-cherry flower and leaf* **RIGHT:** *Giant dogwood flower*

Harlequin Glorybower
Clerodendrum trichotomum

Barely clearing the bar for inclusion, this little tree is found as a street tree or a small yard tree on the west side of the Cascades. Native to East Asia, it's one of about 250 species, mostly in the tropics. It's not invasive, so you'll only find it where planted.

Family: Lamiaceae

Small tree up to 20 feet (6 meters) tall, developing small, rounded crown, reminiscent of eastern flowering dogwood. **BARK:** Light gray to beige, very rough, warty on older trees, almost corky. **TWIGS:** Pubescent, brown, with large, warty lenticels; buds tiny, triangular, sitting above large, rounded leaf scars that look like Pac-Man. **LEAVES:** 4–9 inches (10–23 centimeters) long, ovate but with overall triangular appearance; long, tapering tip and flat (truncate) or wedge-shaped (cuneate) base; margin entire, rarely with small serrations; dark green on top, while leaf underside and petiole are *pubescent; strong smell of peanuts when crushed.* **FLOWERS:** White, fragrant, showy, with 5 petals fused into a tube for half their length; long stamens extend out; appear in late summer in flat-topped clusters. **FRUIT:** Electric-blue drupe subtended by bright magenta flower calyx, a very striking pair.

SPECIES REMARKS: This species tends to have something for each season to help identify it: the peanut scent in spring and summer; the flowers and fruit in late summer and fall; and the fuzzy twigs with oppositely arranged buds in the winter. Few trees will be confused with it.

153

Crepemyrtle
Lagerstroemia spp. hybrids

"*Lagerstroemia* spp. hybrids" is as far as I'm willing to break down these little trees and/or shrubs, spelled varyingly crepe-myrtle and crapemyrtle in other guides. They have been crossed and hybridized beyond imagination, with over one thousand cultivars, making it nearly impossible to apply a species name to an individual tree. I'll leave it to other guides to wade through that and instead focus on the genus. **Family: Lythraceae**

Small trees up to 25 feet (8 meters) tall, upright habit with smooth, slightly curvy, sinuous branches creating tropical look; very often multistemmed. **BARK:** Smooth and beautifully exfoliating, usually creamy beige to cinnamon brown with patches of green or gray in camouflage pattern. **TWIGS:** Slender, angular, with small, corky wings; often reddish; buds small, bluntly pointed, appressed to twigs; *mostly oppositely arranged* but tend to become alternate near end of twigs. **LEAVES:** About 2.5 inches (6 centimeters) long, simple, glossy green, hairless, with very small petioles (almost sessile). **FLOWERS:** Very showy, appear in late summer; purple, pink, red, or white, in terminal firework-like panicles at the edge of canopy; each flower has fringed petals; *very pretty and unlike anything else.* **FRUIT:** Round, dry capsule in same panicle as flowers; 6 valved, opening but persistent through winter.

SPECIES REMARKS: Everyone wants to plant this tree on the warmer west side of the Cascades because it has pretty flowers and it's small. They are gorgeous—I can't argue with that. But don't plant it where a bigger tree that will contribute something to society could go, or at minimum, plant both.

Katsura
Cercidiphyllum japonicum

Another commonly planted street tree that probably shouldn't be, katsura is native to East Asia and does relatively well here but does not appreciate extended drought. You'll find it as a street tree, or less commonly as a yard or park tree, mostly west of the Cascades. **Family: Cercidiphyllaceae**

Medium tree up to 60 feet (18 meters) tall, with upright, pyramidal canopy. **BARK:** Utility gray that develops shallow, criss-crossing ridges with age, revealing orange tones between ridges. **TWIGS:** Slender, brownish-red, usually very straight and swollen where buds emerge; *buds opposite and alternate*; buds red, hooflike, incurved toward twig; on older twigs, buds sit at top of tiny spur shoots. **LEAVES:** Small, blade usually no longer than 2 inches (5 centimeters); heart shaped with crenate or serrate margins; veins radiate from base of leaf and curve toward tip. **FLOWERS:** Dioecious, inconspicuous for both male and female trees; appear in early spring, like tiny feather dusters. **FRUIT:** Small capsule, up to ¾ inch (2 centimeters) long; looks like strange growths at nodes.

SPECIES REMARKS: The buds and leaves are unique in and of themselves, but their opposite or subopposite arrangement will give them away quickly. Any plant that looks similar to the katsura in regard to leaves or size will be starkly opposite or alternate, but not both. Its leaves can be confused with redbud leaves, but those are never oppositely arranged.

155

Chinese Fringetree
Chionanthus retusus

Fringetrees are popular due to their profuse blooms in late spring. Although it is mostly a genus of tropical trees, two species (Chinese fringetree and American fringetree) can be found planted in our region, but mostly in milder areas. Both have opposite or subopposite leaf arrangements, but lean opposite most of the time.
Family: Oleaceae

Small tree up to 25 feet (8 meters) tall, with rounded, bushy canopy. **BARK:** Smooth, reddish to bronze with prominent lenticels; peels away as stems age, then becomes furrowed, dark gray. **TWIGS:** Gray to light brown, swollen where buds are; sparsely endowed with lenticels; buds gray to light brown, like small pyramids, with imbricate scales and slight pubescence. **LEAVES:** Like buds, are opposite or sub-opposite, often along the same twig; 3–8 inches (8–20 centimeters) long, elliptical to obovate, but also sometimes rounded, often appearing (to me) paddlelike; tips either acute or sometimes with a notch (emarginate), margins otherwise entire. **FLOWERS:** Dioecious, white, profuse, emerging in mid- to late spring on erect panicles *borne terminally on new shoots*; petals long and strappy, creating airy, poofy display. **FRUIT:** Blue drupe, ripening in fall; only produced on female trees.

SPECIES REMARKS: The fruit in the fall and the striking bark help to differentiate this species when it's not in bloom.

European Olive
Olea europaea

This classic olive of the Mediterranean is found more often in our region as of late because of its tolerance of drier conditions. Though still fairly uncommon, you can find it planted in yards and along streets, and it will surely become more prevalent over the coming decades. **Family: Oleaceae**

(less than 1 inch, or 2.5 centimeters) drupe; classic olive green that matures to purple with tiny white dots.

SPECIES REMARKS: The thin, gray-green, oppositely arranged leaves sets the olive apart. It may be confused at first glance with a few other evergreen species or Russian-olive, but those have alternately arranged leaves (see table 21).

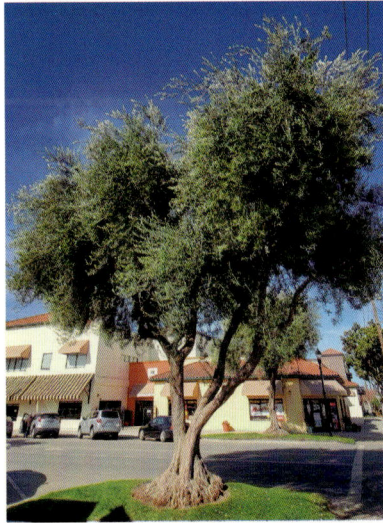

Small evergreen tree up to 30 feet (9 meters) tall, with somewhat messy, rounded crown; tends to look like an out-of-control poof for first several years, but shapes up nicely with age. **BARK:** Rough, somewhat fibrous looking, gray, maturing to be very gnarly and bumpy, very attractive on old trees. **TWIGS:** Spindly, light green, but glaucous gray on youngest shoots. **LEAVES:** *Evergreen*, lanceolate, up to 3 inches (8 centimeters) long, rough to the touch (scurfy); gray-green on top, whiter and glaucous below; veins not obvious. **FLOWERS:** Small, fragrant, appearing in dangling clusters in leaf axil; can be hard to see. **FRUIT:** The olive, a small

Japanese Tree Lilac
Syringa reticulata

A small tree from East Asia, Japanese tree lilac is commonly found as a street tree on the west side of the Cascades where only a small tree can go, like below power lines or in very small planting strips. It's not common in yards or parks and isn't invasive, so the street is the best place to find it. **Family: Oleaceae**

Small tree up to 25 feet (8 meters) tall, with rounded crown; often appears half-dead (to me) because of harsh growing conditions and tendency to grow more like a big shrub. **BARK:** Maintains youthful, shiny appearance, ranging from grayish-purple to polished bronze; uniformly covered in obvious horizontal lenticels, usually with outermost layer peeling away. **TWIGS:** Stout, olive green to bronze, with pronounced, warty lenticels that stick around as the tree ages; buds yellow, egg shaped, with many imbricate scales; often paired at tip of twig. **LEAVES:** Usually

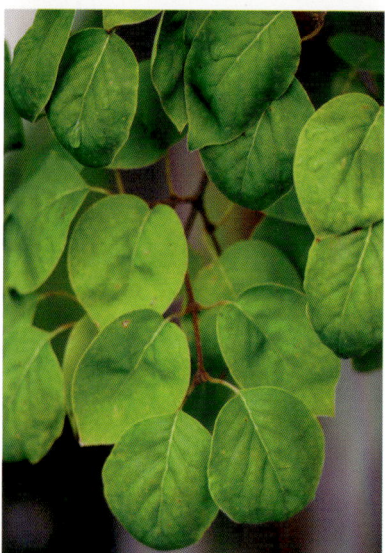

CLOCKWISE FROM TOP LEFT: *Japanese tree lilac fruit and twig; Chinese tree lilac leaves; Japanese tree lilac leaves*

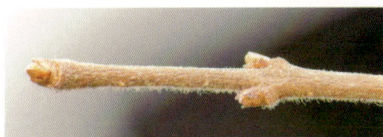

no more than 5 inches (13 centimeters) long, ovate with a pointed tip; often arched downward but with a V shape caused by the blade bending up on either side of midvein. **FLOWERS:** Large terminal panicles of small, white, fragrant flowers that appear in late spring or early summer; profuse when in bloom, resembling its shrubby cousin, the common lilac (*S. vulgaris*), but with bigger panicles of smaller flowers. **FRUIT:** Green, short, paddle-shaped capsule that dries to brown and splits open.

SPECIES REMARKS: The large flower and fruit clusters set this species apart from similar trees with oppositely arranged leaves. You may find **Chinese tree lilac** (*Syringa pekinensis*) listed in some sources, but it has been relegated to a subspecies of Japanese tree lilac: *S. reticulata* subsp. *pekinensis*. Its leaves are smaller and rounder, its twigs are fuzzier, and its bark is more olive green with intensely exfoliating orange-bronze outer layers (see photo).

TOP LEFT: *Japanese tree lilac flower*
TOP RIGHT: *Chinese tree lilac flower, twig, and bark*

ABOVE: *The flowers of American fringetree grow from the previous year's growth, not terminally as Chinese fringetree does.*
RIGHT SIDE, TOP TO BOTTOM: *Seven sons flower flowers, leaves, fruit, and bark*

Similar Species

American or white fringetree (*Chionanthus virginicus*) is also found in the landscape and can be hard to tell apart from Chinese fringetree if not in bloom. Its leaves are slightly larger and not emarginate, and its flowers bloom from *the previous year's growth* and not terminally, a surefire tell between the two species.

Seven sons flower (*Heptacodium miconioides*) is becoming more popular on the west side of the Cascades and could be confused with Japanese tree lilac. Both have terminal clusters of white flowers and oppositely arranged leaves (its family is Caprifoliaceae), but seven sons flower has longer, more lanceolate leaves with *three prominent main veins*. They also tend to hang downward, almost lazily, off the twigs.

Its fragrant flowers don't come out until late summer, well after the tree lilac's, and they don't quite blend together into big cloudlike bursts but rather stay quite independent of each other. They then morph into bright red, sepal-enclosed fruit that just looks like a bright red flower, nothing like the capsules of the lilacs. Finally, the bark is fibrous, light brown, and peels off in long vertical strips, appearing quite stringy.

CLOCKWISE FROM BOTTOM LEFT: *Northern catalpa flower; northern catalpa form; northern catalpa mustache; southern catalpa fruit*

CATALPAS

Catalpa

Some of my favorite trees, catalpas (family Bignoniaceae) put on a gorgeous flower display in the spring and grow in unique and curious forms. For a nonnative, noninvasive species, they are surprisingly common outside of larger cities, often found near old farms or town parks throughout the region. Only two species are commonly found in the Pacific Northwest, plus two hybrids that are getting more traction.

Catalpas have big, floppy, oppositely arranged leaves; large panicles of white flowers; and long, pea pod–like fruit that hangs on all winter (the fruit looks like string beans to some, giving rise to the name "string bean tree"). Inside the fruit are tiny winged seeds that I lovingly call mustaches. Planted purposefully and not found in natural areas, they are yard and park trees, and often street trees in bigger cities. In the eastern portion of the region, cities more often report northern catalpa (*C. speciosa*), while the west side reports more southern catalpa (*C. bignonioides*), but both are likely present in both areas, so check the details to be sure of your identification.

TABLE 9. QUICK GUIDE TO CATALPAS

SPECIES AND KEY TRAITS	SEED AND LEAF
Northern catalpa *Catalpa speciosa* Biggest leaves, up to 12 in. (30 cm); densely hairy below and along petioles; scentless when crushed; bark thick, gray, furrowed; blooms earliest; widest mustaches, with more uniformly combed hairs	
Southern catalpa *Catalpa bignonioides* Leaves up to 10 in. (25 cm), with 1–2 "lobes"; slightly hairy below on veins only; nutty scent when crushed; bark gray with orange, thin, spiraled on stems; blooms 2 weeks after northern, mustaches thinner with tufted ends	
Chinese catalpa *Catalpa ovata* Smallest leaves, usually less than 10 in. (25 cm) and with two "lobes"; mostly hairless at maturity, minimally scented; bark gray, rough; blooms at same time as southern but flowers are a dull yellow; seeds very small	

SPECIES AND KEY TRAITS	SEED AND LEAF
Purple catalpa *Catalpa × erubescens*	
Leaves similar to southern's, but emerge dark purple in spring, often retain darker tones; bark similar to southern's; flowers large, profuse like southern's, but more yellowish like Chinese catalpa; seeds intermediate between parents	

Catalpas have dozens of mustachioed seeds packed into long skinny pods. From top down: northern, southern, and Chinese

Northern Catalpa
Catalpa speciosa

The biggest of our catalpas, northern catalpa is native to eastern North America. Its name is derived from the Indigenous name for the tree (*kutuhlpa*), and the specific epithet means "showy," referring to its lovely flowers.

Large tree up to about 60 feet (18 meters), taller than it is wide; architecture very upright, especially compared to southern catalpa; gets big. **BARK:** Gray, rough, furrowed, thickest of region's catalpas. **TWIGS:** Very stout, rounded, tan, with pronounced lenticels and circular leaf scars, with ring of vascular bundles in middle; buds small, dome shaped, set right atop the leaf scar. **LEAVES:** *Opposite or in whorls of 3*, large, 6–12 inches (15–30 centimeters) long, ovate with an acuminate tip and heart-shaped or flat base; potentially 1 or 2 points along the margin that look like beginnings of lobes; glabrous above, *densely pubescent below* and along the petiole; *scentless* when crushed. **FLOWERS:** Tubular, white with purple dots and yellow smudges along the bottom; large, showy, upright panicles like pom-poms; first to flower in early summer. **FRUIT:** Long, skinny pod, superficially like a pea pod, but rounded (not flat) and filled with little seeds that look like mustaches.

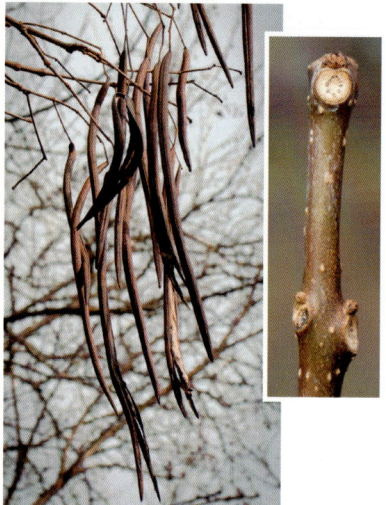

SPECIES REMARKS: The leaves and pods make this tree stand out almost any time of year; a closer look at the buds or leaves, in whorls of three on thick stems, makes for a sure identification. It is easily confused with southern catalpa and the hybrids, so look closely at the seeds and pubescence to distinguish them.

Southern Catalpa
Catalpa bignonioides

The smaller of our two common species, southern catalpa can be difficult to tell apart from northern catalpa. It's often found in yards, parks, and as a street tree, and I get the feeling a few trees around the landscape seeded themselves there.

SPECIES REMARKS: The minute details are what you need to tell the catalpas apart, specifically their bark, form, and the hairiness of the leaf underside. The seeds are the most consistent trait I've found to separate them; you can usually find some in old pods on the tree year-round.

Medium tree up to about 40 feet (12 meters) tall with a lower, more spreading canopy. **BARK:** Light gray with orange tinge in shallow furrows, often appears spiraled on stems; thinner than northern catalpa's. **TWIGS:** Similar to northern, light brown with distinctive oval, cup-like leaf scars; buds are dome shaped, perched just over leaf scar. **LEAVES:** Look very similar to northern but are only about 10 inches (25 centimeters) long; below, they are slightly pubescent on the leaf veins only (not down the petiole); *strong nutty odor when crushed.* **FLOWERS:** Similar to northern, but bloom two weeks later and are considered by many to be more profuse. **FRUIT:** Similar to northern's but smaller, not more than ½ inch (1 centimeter) wide, with thinner walls; seeds are also thinner, with more tufted hair on ends (not as widely combed as northern).

Similar Species

Chinese catalpa (*Catalpa ovata*) isn't all that common, but a hybrid between it and southern catalpa called **purple catalpa** (*Catalpa ×erubescens*) is gaining popularity. Chinese catalpa has smaller leaves that tend to have three pointed "lobes" (arguably just raised points along the margin; see top photo above). They emerge deep purple in spring, starting off pubescent and then becoming almost entirely glabrous at maturity. Its flowers are smaller and dull yellow, with smaller pods that have equally tiny mustache seeds. Purple catalpa has traits intermediate between its parents: profuse but yellowish flowers and larger, purple-tinged leaves with three "lobes" and pubescence on their undersides along the veins (see bottom left photo).

Chitalpa (× *Chitalpa tashkentensis*) is a hybrid between southern catalpa and desert-willow (*Chilopsis linearis*) and exhibits traits of both. Its long, lanceolate leaves have acuminate tips (like an arrowhead), and they're opposite as often as they're whorled or alternate (see right photos above). Check multiple twigs to get a good sense, and if you see this variation, that's a good clue it's this species. The flowers are a light pink version of southern catalpa's, but they're sterile, so you won't find any fruit. Overall, it's a small tree, not growing more than 25 feet (8 meters) tall, with smooth bark and a rounded, airy crown. They tend to have poor root systems, especially when getting established, so they often end up with crooked stems.

Paulownia
Paulownia tomentosa

Paulownia, also known as empress tree, exhibits the same unsavory invasive traits as the tree of heaven, which has landed it on naughty lists across the region. Growing many feet in a year, often becoming a menace, it is readily found in yards, ignored margins of the landscape, and cracks in the sidewalk. **Family: Paulowniaceae**

Medium tree up to about 40 feet (12 meters) tall, with rounded canopy and often dominant central leader for much of its lower heights. **BARK:** Gray, smooth when young, developing vertical striations that eventually split into isolated, but closely spaced, warty plates. **TWIGS:** Stout, light brown, with prominent lenticels; leaf scars round, like craters; buds small, domed, sitting atop edge of leaf scar. **LEAVES:** Oppositely arranged, massive, nearly 12 inches (30 centimeters) long and wide; on vigorous shoots they can be even larger; mostly heart shaped (cordate), especially at base, with entire margins, but sometimes with 3 distinct points that give it a house-like shape; *lower* side densely pubescent. **FLOWERS:** Gorgeous violet-purple; emerge before leaves on large, pubescent, burnt-orange panicles; flower buds are set in summer and can be seen all winter. **FRUIT:** Oblong, woody, beaked capsule that splits open in fall, releasing dozens of tiny, winged seeds; capsules often persist through winter with seeds still inside. **SPECIES REMARKS:** The big, fuzzy, floppy leaves will set this species apart from others in summer, and the capsules will give it away in the winter if they are still present. It looks most like the catalpas, but they have long pod fruit and white (not purple) flowers.

CLOCKWISE FROM LEFT: *Bigleaf maple leaf; sycamore maple samara; Norway maple form*

MAPLES

Acer

Oh, the maples (family Sapindaceae)—they have one of the most ubiquitous and recognizable leaves, probably second only to the oaks, and even then, it would be close. Maples are extremely common in the landscapes of the Pacific Northwest, and despite the region having only three native species (potentially two others at a stretch), inventories of street trees show that maples often make up more than 25 percent of the total population.

Though the "maple leaf" shape is very well-known, it can often be a poor indicator of an actual maple. For example, David's maple (*Acer davidii*) is a part of a group called the snakebark maples, which almost all have simple, unlobed, and seemingly pinnately veined leaves. Two species of maple are listed at the start of the next section ("Opposite, Compound, Pinnate") because they have pinnately compound leaves, which means they don't look very maplelike. Conversely, trees like sweetgums and sycamores have leaves that scream maple, but they are in fact unrelated.

To tell if you're looking at a maple, a pair of traits will give it away every time: the buds and leaves will be oppositely arranged, and the fruit will be a double samara. If you see both traits, then the next question is, which maple is it? No other trees have double samaras, and each of the maple species in the region has its own unique style. The common Northwest species also differ substantially in their leaves, buds, form, and bark. So, if no samaras are present, pick a different trait and eliminate species that don't match it, then move on to another to further narrow the field. You'll almost always find two good traits to use year-round.

TABLE 10. QUICK GUIDE TO MAPLES

SPECIES & KEY TRAITS	LEAF	SAMARA
Norway *A. platanoides* Bristle-tipped leaves and milky sap in petiole; dark gray bark with crisscrossing ridges		
Sugar *A. saccharum* Round-tipped leaves, no milky sap; light gray, platy bark		
Bigtooth *A. grandidentatum* Leaves like sugar's, but with deeper sinuses and lobes that are even more rounded		
Sycamore *A. pseudoplatanus* Wrinkly leaves; patchy, orange and gray bark; samaras on long racemes		
Hedge *A. campestre* Small leaf, rounded lobes, milky sap in petiole; sometimes corky wings on twigs		

SPECIES & KEY TRAITS	LEAF	SAMARA
Trident *A. buergerianum* 3 obvious, sharp lobes; shiny, beige bark, exfoliating in vertical strips		
Red *A. rubrum* Usually 3 lobes with a rounded base, not shiny; red flowers in early spring; gray, platy bark		
Silver *A. saccharinum* Deep sinuses, sharply serrated lobes; bark dark gray, very flaky		
Freeman *A. × freemanii* Usually very columnar growth; leaves and bark a mix between red and silver maples, its parent species		
Bigleaf *A. macrophyllum* Leaves biggest of all common species, with deep sinuses, overlapping lobes; samaras have sharp hairs on seed		
Japanese *A. palmatum* Leaves very small, with varying sinus depth, lacey appearance; often purple leaves		

SPECIES & KEY TRAITS	LEAF	SAMARA
Fullmoon *A. japonicum* Very similar to Japanese maple, but leaves have more lobes that are bigger, longer, more deeply incised		
Vine *A. circinatum* Shrubby; leaves with extremely shallow sinuses; samaras often held above leaves when maturing in late summer		
Rocky Mountain *A. glabrum* Mostly shrubby or small; leaves in 3 serrated lobes or trifoliate farther east		
Amur or Tatarian *A. tataricum* Dense, shaggy canopies with profuse samaras that stay all winter; note two leaf types depending on subspecies		
Paperbark *A. griseum* Leaves trifoliate, pubescent; bark orange, exfoliating		
Boxelder *A. negundo* Pinnately compound leaves; long chains of hooked samaras		

Norway Maple
Acer platanoides

This is the most common maple in the Pacific Northwest, bar none, so it can be found almost anywhere in the landscape, including in some natural areas and ignored landscape margins on the west side of the Cascades, as it's slightly invasive.

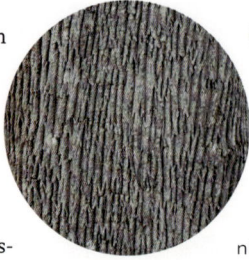

Medium tree up to 60 feet (18 meters) tall and wide; does not maintain a central leader, but instead develops large scaffold limbs with a globe-like canopy. **BARK:** Dark gray with shallow, crisscrossing ridges, similar to bigleaf maple. **TWIGS:** Smooth, hairless (glabrous), stout, with single, plump terminal bud between 2 smaller accessory buds; bud scales imbricate, reddish-green or entirely green. **LEAVES:** 4–7 inches (10–18 centimeters) long, palmately lobed with 5 prominent lobes *ending in bristle tips*; appears wider than long, with particularly long petiole that *exudes milky sap* when removed from stem. **FLOWERS:** Greenish-yellow, appear before leaves in early spring in dense clusters (corymbs). **SAMARAS:** Each up to 2 inches (5 centimeters) long, wings spreading away from each other; glabrous, green fading to yellow-brown in midsummer (see table 10).

SPECIES REMARKS: Eighty-nine cultivars of Norway maple have been reported, many of which are planted in the Pacific Northwest, including purple-leaf, variegated, upright, and compact varieties and all manner of combinations.

Sugar Maple
Acer saccharum

The classic maple syrup tree of eastern North America, sugar maple can be found most often as a street tree in the Pacific Northwest and only sometimes as a park or yard tree. It's far less common than the easily confused Norway maple, and it's not found in natural areas.

Large tree up to 75 feet (23 meters) tall, with upright, rounded canopy at maturity; tends to maintain a central leader with many upright lateral limbs creating very uniform branch architecture. **BARK:** Brownish-gray, smooth when young, developing long, hard ridges and plates that separate away from trunk with age. **TWIGS:** Reddish-brown, slender, straight, glabrous, with lenticels; buds *dark brown* to slightly reddish, with a single *sharply pointed* terminal bud and 2 smaller accessory buds on either side. **LEAVES:** Palmately lobed with 5 lobes (rarely 3); similar to Norway maple, but *tips are rounded, lack bristles*, sinuses are deeper, and it *does not exude milky sap* from petiole (see table 10). **FLOWERS:** Yellow-green, appear before leaves in early spring; not showy or wildly helpful for identification. **SAMARAS:** Around 1 inch (2.5 centimeters) long, each with very smooth, rounded base; both wings grow down, forming a horseshoe shape; mature in summer.

SPECIES REMARKS: Sugar maple is planted widely and has a lot of cultivars, but they all tend to maintain their telling leaf, twig, and samara characteristics. In winter, look for the upright, oval habit and dark brown, pointed buds; in summer, the rounded (not bristle-tipped) lobes and horseshoe-shaped samaras with very rounded seed bases set it apart.

SUGAR OR SUGAR-FREE?

There are at least four other maples that are variously described as subspecies or varieties of sugar maple or as full species in their own right: black maple (*Acer nigrum*), chalk maple (*A. leucoderme*), southern sugar maple (*A. barbatum*), and **bigtooth maple** (*A. grandidentatum*). They all appear to be very closely related to sugar maple proper, and they not only look similar but also tend to interbreed where their ranges overlap. Only bigtooth maple is widely accepted as its own species, while the others are considered subspecies of sugar maple.

Lucky for us, bigtooth maple is the only one of these four that you're likely to find in the Pacific Northwest's landscapes with any real frequency. It differs from sugar maple in its leaves, which are a little smaller with deeper sinuses, and every lobe or tooth coming off a lobe is much more rounded. Bigtooth maple's samaras also tend to grow more out than down, though this can be somewhat variable. The 'Rocky Mountain Glow' cultivar is the most common.

TOP & MIDDLE: *Bigtooth maple samara and leaf*
BOTTOM: *Sugar maple leaf*

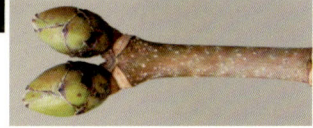

Sycamore Maple
Acer pseudoplatanus

Widely planted (and semi-invasive) in the Pacific Northwest, this European native can be found in yards, parks, and along streets, as well as in marginal spaces where it has seeded itself in. It's most commonly found west of the Cascades.

Large tree up to 60 feet (18 meters) tall, with rounded crown spreading up to 40 feet (12 meters). **BARK:** Light gray with light orange exposed as outer bark falls away; exfoliates in squarish plates when young, but more rounded plates with age, creating mottled, 2-tone appearance. **TWIGS:** Light reddish-brown with many lenticels; buds green and plump, with imbricate scales, like Norway maple but remaining green. **LEAVES:** Palmately lobed, blade 4–7 inches (10–18 centimeters) long and wide with mostly 5 lobes (sometimes 3); dark green above, lighter (sometimes reddish) below; coarsely toothed margins and prominent veins, giving unique ruffled texture and appearance; petiole as long as leaf blade. **FLOWERS:** Yellow-green, appear *after* leaves in long racemes (not corymbs, unlike Norway or sugar maple). **SAMARAS:** About 1.5 inches (4 centimeters) long, borne in pendulous racemes; angle between wings about 60 degrees; turn showy reddish-orange in late summer through fall, usually in great numbers on tree.

SPECIES REMARKS: Being invasive, like Norway maple, sycamore maple is often found growing in places ill-suited for a large tree. The most common cultivar, 'Atropurpureum', has *purplish-red leaves and usually three lobes*. The bark, pendulous racemes, and wrinkly, serrated leaves set this species apart.

Hedge Maple
Acer campestre

This commonly planted European species, also called field maple, should earn a Cutest Maple Leaves award, in my opinion. You'll find it mostly along streets and in parks, but also popping up in nearby vacant beds, though it doesn't cause any issues.

blackish, with imbricate, slightly pubescent scales. **LEAVES:** Palmately lobed with 5 *rounded lobes* (sometimes 3); blade generally about 3–4 inches (8–10 centimeters) long and wide, with similar-length *petiole that produces milky sap* when broken. **FLOWERS:** Green, inconspicuous, appearing with leaves in clusters (corymbs). **SAMARAS:** Each about 1 inch (2.5 centimeters) long, appearing in late spring and maturing to light brown over summer; nearly horizontal. **SPECIES REMARKS:** Hedge maple leaves can appear like smaller versions of bigtooth maple's, but its samaras are not similar at all, and they grow much more slowly and stay much smaller. A somewhat old-fashioned tree, it is infrequently planted today but is a great choice for tough situations.

Medium tree up to 45 feet (14 meters) tall, with uniformly rounded crown, and handsome branch architecture; slow growing, developing a compact, dense canopy with soft appearance. **BARK:** Uniformly dark gray, often broken up into small, hard, rectangular blocks that can fall away, leaving lighter gray bark below; trunk tends to develop slightly twisted, ripply appearance. **TWIGS:** Slender, short (slow grower), light brown to brownish-orange, sometimes with corky growth; buds small, dark brown to

Trident Maple
Acer buergerianum

Native to China, Taiwan, and Japan, trident maple is found most often planted as a street tree, especially in small planting spaces or below power lines. Its leaves don't appear very maplelike at first glance, but luckily it's often laden with samaras.

Small tree up to 30 feet (9 meters) tall but usually shorter, with rounded, dense, but shaggy crown. **BARK:** Light gray, becoming darker with age, but soon *exfoliating in small, scaly plates*, revealing light orange bark underneath; plates usually somewhat vertically oriented. **TWIGS:** Slender, reddish-brown, *pubescent*, though becoming gray-brown, glabrous with age; buds small, reddish-orange, with striking imbricate scales fringed with white hairs; pointed, terminal bud has 2 accessory buds at end of twig. **LEAVES:** Up to 3 inches (8 centimeters) long; split into *3 distinct, shallow lobes* (hence its name) with triangular (cuneate) or rounded base; margin entire or softly serrated; upper side lustrous green; often shaggily overlapping each other. **FLOWERS:** Inconspicuous; come out with leaves in early spring in dense clusters. **SAMARAS:** About 1 inch (2.5 centimeters) long with *tight angle* between wings; mature in late summer, often profuse on tree through fall.

SPECIES REMARKS: Leaves are comparable to red maple's, but lobes are far more distinct, with smooth margins, a glossy surface, and no more than three lobes (red maple can sometimes have five, including two tiny lobes near the leaf base).

Red Maple
Acer rubrum

Found along almost any street and in front yards, this is one of the most common trees planted anywhere in North America. It's a nursery trade juggernaut too, so there's likely a cultivar or hybrid for every day of the year, including leap years. You'll only find it where it has been planted, not in natural areas.

Medium tree, usually up to 50 feet (15 meters) tall, with rounded crown about as wide. **BARK:** Light gray, smooth when young, then developing ridges and lifted plates; appearance in between that of sugar maple and silver maple, but not as thickly plated or shaggy as either. **TWIGS:** Skinny, red, straight, often with many short shoots as you go down; vegetative buds small, red, tightly appressed to twig; floral buds spherical, clustered around nodes, similar to silver maple but not as pointed. **LEAVES:** Palmately veined with 3 *prominent, shallow lobes* (often 2 more tiny lobes near base) and long petiole; blade usually 3–4 inches (8–10 centimeters) long with medium serrated margins; green on top, lighter below; vibrant red in fall. **FLOWERS:** Red, appearing in dense, pendulous corymbs before leaves in early spring (one of the first trees to bloom). **SAMARAS:** Very small (less than ¾ inch, or 2 centimeters, long), red, appearing swiftly after flowers in early spring, usually fallen away by June; nearly horizontal.

SPECIES REMARKS: The roots of red maple will form solid masses at the base of the tree when it's planted in a contained space like a tree well. The rounded floral buds and short shoots make winter identification a breeze, and in summer, the three-lobed leaves are the best clue.

Silver Maple
Acer saccharinum

Silver maple is found across our region, especially on the east side of the Cascades. Though not invasive, silver maple seems to do well in harsher soils, so you'll often find it growing where no other trees can make it.

SPECIES REMARKS: Bark and leaves are key characteristics for this species. No other maples—really no other trees broadly—in the Pacific Northwest have such deeply incised leaves with the contrasting green and white color *and* such shaggy bark.

Large tree up to 80 feet (24 meters) tall, with spreading crown up to 50 feet (15 meters) wide; develops large, straight scaffold limbs that grow up and out from canopy at precarious angles, usually lacking central leader early on. **BARK:** Gray, smooth when young, developing small, lifted plates that exfoliate in vertical strips, most prominently along lower stem. **TWIGS:** Red to reddish-brown, straight, with lenticels; buds similar to red maple, though slightly more pointed. **LEAVES:** Palmately veined with 5 lobes separated by *deep sinuses*; blade 4–6 inches (10–15 centimeters) long, *sharply toothed, incised along margin*; dark green on top, silvery white below, highly contrasting. **FLOWERS:** Yellowish with red stamens, appearing before leaves in messy clusters. **SAMARAS:** Up to about 2 inches (5 centimeters) long, wrinkly, paddle shaped; mature to tan in early summer; angle between wings 90 degrees or less.

THE MYRIAD MAPLES

Red maple and silver maple are closely related species, and dozens of cultivars of each have been developed for the nursery trade. They have also been hybridized extensively, creating the hybrid **Freeman maple** (*Acer × freemanii*). These hybrids have been backcrossed with the species to create still new varieties, which makes identification down to a species or cultivar quite challenging, requiring consideration of their height, fall color, leaf shape, overall form, and even when they were planted (see table 10 for leaf and samara photos).

Armstrong and Autumn Blaze red maples (both varieties of Freeman maple) are very common and have traits that are very much blends of the parents, most notably leaves with three lobes (like red maple) but that are deeply incised (like silver maple) and bark that looks more like silver maple but that does not flake away. As with other hybrids, if you see a maple that does not neatly fit into either silver maple or red maple, you almost certainly are dealing with a cultivar, hybrid, or any mixture of the two. This is especially true if the mature form of the tree appears to be notably upright, a classic trait of Armstrong red maple.

Bigleaf Maple
Acer macrophyllum

The giant native species of the Pacific Northwest, this maple can be found growing mostly on the west side of the Cascades, both in urban areas and very commonly in natural areas. It tends to grow naturally in ignored spots, like ditches, hillsides, or vacant lots, but it can be a forest tree in its own right.

Large tree up to 100 feet (30 meters) tall, but more commonly up to 80 feet (24 meters); crown large, vase shaped, spreading, with central leader often splitting into a few main scaffold limbs that arch up, then outward. **BARK:** Smooth, gray with green striations when young, developing cross-hatched ridges uniformly covering stems with age. **TWIGS:** Stout, reddish-green, straight on vigorous shoots, more irregularly curved and branched on less vigorous shoots; obvious lenticles; buds large, red, with imbricate scales; usually single terminal bud flanked by 2 large accessory buds. **LEAVES:** Huge, 8–12 inches (20–30 centimeters) long, sometimes larger; pal-mately lobed with 5 main lobes and deep sinuses; 3 central lobes tend to overlap one another; biggest of Pacific Northwest maples; *petiole exudes milky sap when broken from stem.* **FLOWERS:** Yellow to green; appear before leaves in early spring on long, drooping racemes; very striking, covering whole tree. **SAMARAS:** Hang in pendulous racemes like flowers, 1.5 inches (4 centimeters) long with 60-degree angle between them; *base covered in sharp hairs* that will stick in your skin if touched, only maple in region to have this trait.

SPECIES REMARKS: Older trees are covered in thick mats of moss and licorice fern, which is a consistent identification characteristic; few other trees in our region have so much epiphytic growth.

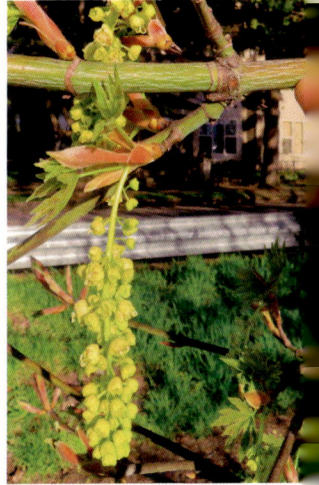

Japanese Maple
Acer palmatum

A favorite tree for people who don't want a tree. Japanese maple has a huge amount of genetic variation that has produced an unending supply of cultivars. It's very commonly planted in yards, less frequently as street trees or in parks, and not found in natural areas.

Small tree up to 25 feet (8 meters) tall, with round crown; spreading, curvy limbs; delicate texture. **BARK:** Gray, smooth, developing few shallow furrows with age; tends to have striations running down stems and rippled, sinuous appearance (some cultivars have greenish or reddish bark). **TWIGS:** Slender, reddish to light orange or greenish, with 2 red or green buds at terminal end; buds small, incurved; nodes somewhat flared, with buds appearing to sit on pedestals. **LEAVES:** Small; blade usually not more than 3 inches (8 centimeters) long; palmately lobed with *5–7 (up to 9) triangular lobes*; serrated margins; deep sinuses, often with lacy appearance on cutleaf varieties. **FLOWERS:** Red, inconspicuous, appearing with leaves in early spring. **SAMARAS:** Small, about 1 inch (2.5 centimeters) long each; maturing in late summer, wings flare out just shy of horizontal, looking like a fake mustache; often maintains reddish color, especially purple-leaf varieties.

SPECIES REMARKS: Japanese maple has been cultivated for thousands of years; as many cultivars are available as anyone could ever want. The normal species has green leaves, but more often you'll find a purple-leafed variety with normal leaves or deeply cut sinuses, so much so that they appear like strings of lace. There are also weeping or dwarf varieties and those with multicolored bark, so it's important to understand the variation in this species. Only fullmoon and vine maples (see next pages) are close in terms of form and leaf appearance.

Similar Species

Fullmoon maple (*Acer japonicum*) is easily confused with Japanese maple, being similar in almost all regards, from buds to samaras to its form and architecture. The main difference you'll notice is the leaves: fullmoon maple has slightly larger leaves with *seven to eleven lobes* separated by deep sinuses (*A. palmatum* has only five to seven, rarely nine). The lobes tend to split into more, smaller lobes and overlap near their tips.

Vine maple (*Acer circinatum*) is one of our diminutive native maples from the west side and is planted frequently in landscapes as a small tree, though it's often demoted to "large shrub," which I'm inclined to agree with. They're short, multistemmed from the ground, and have leaves with mostly seven to nine lobes and *very shallow sinuses*. It's only mentioned here because it's often confused with Japanese maple, and I wouldn't hear the end of it if I omitted it entirely.

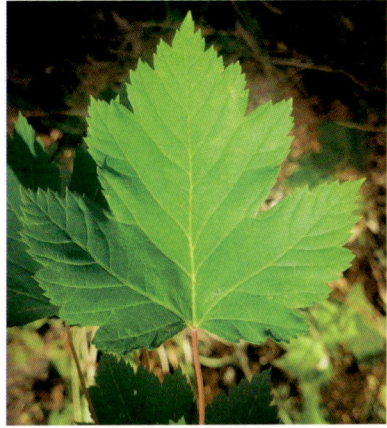

Rocky Mountain and Douglas maples (*Acer glabrum*) are the other diminutive maples native to our region. Douglas maple (var. *douglasii*) is shrubby and grows on the west side of the Cascades, while Rocky Mountain (var. *glabrum*) grows on the east side and is more treelike, with three to five palmate lobes that sometimes split into leaflets. Pay careful attention to the bud arrangement and fruit.

Amur maple (*Acer tataricum* var. *ginnala*) and **Tatarian maple** (*A. tataricum* var. *tataricum*) are two varieties that may throw you off when you see them because they often lack classically maplelike leaves. Previously classified as two different species (Amur was *A. ginnala*), they have leaves that are simple, doubly serrated, and mostly unlobed for Tatarian, whereas mostly three lobed for Amur. Both trees produce showy flowers (for a maple) in rounded panicles that become bright red samaras in summer.

OPPOSITE, COMPOUND, PINNATE

The trees in this section exemplify the Stepwise Journey the best: these are the trees with oppositely arranged, pinnately compound leaves, and in the Pacific Northwest, only three genera that are also firmly considered trees fit these criteria. (The elderberries, in the genus *Sambucus*, are the closest to being a fourth, but they are more shrub than tree, so for the purposes of this guide, they didn't make the cut.) If you were to observe only the bud arrangement and leaf type, you could immediately and confidently narrow down the possible species of all the broadleaf trees in the region to either an ash, one of two maples, or a corktree—from hundreds down to about 12!

Once you've narrowed down your options so substantially, identifying the final species takes only a step or two more down the Stepwise Journey. Several traits are very useful at this point. For example, the corktrees have black berrylike fruit, the ashes have dangling groups of single samaras, and the maples have double samaras; find the fruit, and you've narrowed down your tree to one specific genus. You can also choose another trait to compare, like the number of leaflets, the attributes of those leaflets, the buds, or the bark; each trait you focus on will help you to narrow the field until one tree jumps out. From twelve trees to one, just like that.

LEFT: The oppositely arranged buds of European ash **RIGHT:** The pinnately compund leaf of Chinese corktree

Paperbark Maple
Acer griseum

A very popular ornamental tree native to China, paperbark maple is often planted for its vibrant bark, slow growth, and small mature form. It's found often as a street tree or landscape tree, but never in natural areas. Like the boxelder, the leaves are compound.

Small tree up to 25 feet (8 meters) tall, with rounded canopy, upright growth. **BARK:** Dark brownish-red, turning burnt orange as *outer layers exfoliate* from stem, creating a very striking textural appearance; few other trees even come close. **TWIGS:** Tend to be short due to slower growth, but straight, bronze to brownish-green, with light pubescence; begin to exfoliate in second and third years; buds dark brown to black, pointed, with imbricate scales and tufts of hair along scale edges. **LEAVES:** *Trifoliate* with rounded, sparsely toothed margins; leaflets up to 2.5 inches (6 centimeters) long, shiny green above, paler whitish below with pubescence; petiole and leaflet stems (petiolules) *densely pubescent.* **FLOWERS:** Yellow, inconspicuous, solitary or in clusters. **SAMARAS:** Up to 1.5 inches (4 centimeters) long, with rounded base covered with short hairs, 60–90-degree angle between each; mature in late summer to brownish-yellow.

SPECIES REMARKS: Paperbark maple is hard to misidentify if you look for the opposite leaf arrangement and the orange bark. Only madrone has similar bark, but it's evergreen and has alternately arranged simple leaves, among other differences. Paperbark maple tends to jump out at you once you see it.

185

Boxelder
Acer negundo

Considered native in the Pacific Northwest by some, invasive by others, this predominantly eastern species appears to be expanding its range to the west no matter how you see it. Also known as Manitoba maple, it can be found as a landscape or street tree, but just as often it is found growing wild in ditches or disturbed areas east of the Cascades.

Medium tree up to 40 feet (12 meters) tall, with rounded canopy often as wide as or wider than it is tall; tends to sprout from interior canopy and can develop a messy, irregular habit. **BARK:** Gray, developing ridges and furrows as it ages, becoming rough, often developing burls. **TWIGS:** Covered in whitish bloom (glaucous) that rubs off to reveal reddish-brown beneath; buds light gray, bulbous, onion-like, clustered at tips. **LEAVES:** 6–8 inches (15–20 centimeters) long, *pinnately compound* with 3–5 leaflets (up to 7 or 9); may start out with slight pubescence below but become hairless (glabrous) on both surfaces; leaflet margins variably toothed or smooth (entire); short petiolule; often a variegated version is planted. **FLOWERS:** Dioecious, hanging in pendulous racemes with long peduncles, appearing as long chains. **SAMARAS:** 1.5 inches (4 centimeters) long with skinny wings and inwardly hooked ends; grow at a 60-degree angle at most; mature in late fall and remain on tree through winter; profuse.

SPECIES REMARKS: Boxelder is the most confusing of our common maples due to its pinnately compound leaves, which look very similar to those of ash trees (*Fraxinus* spp.). Look for the bulbous, gray buds and long racemes of hooked double samaras (ash have single samaras that are straight, like an oar).

CLOCKWISE FROM TOP LEFT: *Oregon white ash fruit; Raywood ash form; green ash bark; Raywood twig*

ASHES

Fraxinus

Another big hitter of landscape trees, ashes (family Oleaceae) are widely planted in the Pacific Northwest as street trees. Only one species is native to the region; the rest are mostly ornamental species from eastern North America and Europe. Luckily (or unluckily, as the case may be), only a few species tend to be planted with any regularity.

Ash trees are often confused with other species that have pinnately compound leaves, like the walnuts, hickories, and maackias. But as noted at the beginning of this section, you can separate out ashes quickly by looking for just a few traits. All our ashes have oppositely arranged, pinnately compound leaves and single-samara fruit that hangs in pendulous clusters. Just as with the maples, once you see their telltale traits, the next question is simply, which species of ash is it?

Unfortunately, the invasive insect emerald ash borer (EAB) has finally made it to the Pacific Northwest; it will surely spread throughout the region over the next several years and decades, killing off most of the region's ash species and rendering this section more of a catalog of what once was. So, here's to either hope or nostalgia!

TABLE 11. QUICK GUIDE TO ASHES

LEAF SIZE	LEAFLETS	TWIGS
GREEN *Fraxinus pennsylvanica* | Common		
12 in. (30 cm)	5–9 leaflets (usually 7), ovate, serrated, especially near tip; *glabrous above, pubescent below* but often only on veins	
OREGON WHITE *Fraxinus latifolia* | Common		
12 in. (30 cm)	5–7 leaflets, oval, entire, glabrous above, *quite pubescent below*	
AMERICAN WHITE *Fraxinus americana* | Common		
15 in. (38 cm)	5–9 leaflets (usually 7), ovate, serrations near tip, usually *glabrous above and below*	
EUROPEAN *Fraxinus excelsior* | Common		
12 in. (30 cm)	7–11 leaflets, ovate, coarsely serrated, glabrous, except for sparse hairs along midribs, dense hairs on rachis	
RAYWOOD *Fraxinus angustifolia* subsp. *oxycarpa* | Common		
10 in. (25 cm)	Leaves *whorled in 3s* with 5–7 leaflets, each narrowly lanceolate, sparsely toothed	

LEAF SIZE	LEAFLETS	TWIGS
FLOWERING *Fraxinus ornus* | Semi-common		
8 in. (20 cm)	7–9 leaflets, oval with acuminate tips, *often overlapping*, *terminal* leaflet slightly more rounded	
BLUE *Fraxinus quadrangulata* | Rare		
14 in. (35 cm)	5–11 leaflets, each lanceolate, on short petiolule, sharply serrated, pubescent along lower midvein rachis	
MANCHURIAN *Fraxinus mandshurica* | Rare		
14 in. (35 cm)	9–11 leaflets, oval, sharply serrated; with sunken veins; winged rachis with tufts of orange hairs where leaflets attach	

An American ash in fall can range from yellow to orange to purple.

Green Ash
Fraxinus pennsylvanica

As the most common ash planted in the region, green ash can be found along the streets of almost every city here. It's the most cold hardy and tolerant of poor soils, so it's carved out a good niche for itself from Anchorage to Boise.

Medium tree up to 60 feet (18 meters) tall, with rounded, globe-like canopy; without leaves, often messy looking and twiggy; opposite-branching pattern usually discernable from older, U-shaped branch attachments. **BARK:** Gray, crosshatched, forming a diamond pattern of hard, interlacing ridges uniformly covering stems. **TWIGS:** Rounded, stout, often pubescent, but sometimes lacking hairs; buds brown, woolly, quite plump and spherical compared to other ashes, *set above a semicircular leaf scar*; do not notch deeply into scar (in contrast to white ash). **LEAVES:** Pinnately compound, up to 12 inches (30 centimeters) long, with 5–9 (usually 7) leaflets; each leaflet slightly serrated (mostly toward apex); hairless (glabrous) on top, moderately pubescent below, mostly along veins. **FLOWERS:** Dioecious; small, purplish pollen flowers in tight clusters on male trees; larger, greenish-white seed flowers in looser clusters on female trees. **FRUIT:** Single samara, canoe-paddle shaped, around 1.5 inches (4 centimeters) long, not more than ¼ inch (6 millimeters) wide.

SPECIES REMARKS: Green ash is probably the ash on your street right now, it's that common. The messier canopy sets it apart in the winter, along with the leaf scars and yellow fall color. One cultivar called 'Leprechaun' has tiny leaves and slow growth, which makes for a fun find.

Oregon White Ash
Fraxinus latifolia

This is the Pacific Northwest's only native ash, and it stays mostly west of the Cascades in natural areas or wet backyards. It was planted more in native-focused parks and landscaping before the arrival of EAB, but it never quite broke into the street-tree scene.

Large tree up to 80 feet (24 meters) tall, with upright, somewhat messy canopy; single stemmed or with 2–3 main scaffold limbs; often looks drought stricken in fall. **BARK:** Like other ashes, gray with cross-hatched ridges, creating a diamond pattern. **TWIGS:** Stout, straight, greenish to brown with white lenticels, sparse pubescence; buds rusty brown, woolly like green ash but smaller overall; leaf scar semicircular like green ash, with buds mostly above the flat side. **LEAVES:** Pinnately compound, up to 12 inches (30 centimeters) long, with 5–7 oval, nearly stalkless (sessile) leaflets, often lacking intensely pointed tip or any serrations; *pubescent below*, glabrous above. **FLOWERS:** Dioecious; similar to green ash. **FRUIT:** Paddlelike samara, appearing in late summer in dense clusters on female trees; very similar to green ash.

SPECIES REMARKS: Did I mention it's like green ash? The two are very closely related, so they share many traits. The best clue for this species is context: whereas green ash grows along your street, Oregon ash pops up in the wetland behind your house. The seven rounded, sessile, pubescent leaflets with entire margins separate it on its own merits too, along with its tall, messy habit and smaller buds.

American White Ash

Fraxinus americana

Found without trouble as a street and park tree, American white ash is especially common in newer plantings due to some popular cultivars becoming more available in the west over the last ten years or so. It's not invasive, so you'll only see it where it's been planted.

Large tree up to 80 feet (24 meters) tall in the landscape, usually growing wider than others, with a statelier, cleaner appearance. **BARK:** Usually crosshatched like other ashes, only slightly lighter gray, with slightly thicker, corkier ridges. **TWIGS:** Stout, rounded, light brown to gray, glabrous; buds dark brown (not quite black though), dome shaped, somewhat squat; lateral buds distinctly *set within leaf scar*, giving scar characteristic U or crescent shape. **LEAVES:** Pinnately compound, up to 15 inches (38 centimeters) long, with 7 leaflets (sometimes 5–9), each on a short stalk (petiolule); leaflets usually entire or nearly so and *hairless on both sides*. **FLOWERS:** Dioecious, not wildly helpful in identification. **FRUIT:** Cluster of single, paddle-shaped samaras; about ¼ inch (6 millimeters) wide (slightly wider than green ash).

SPECIES REMARKS: American white ash and green ash may be the only trees that I find easier to tell apart in winter due to their buds and leaf scars. The crescent-shaped leaf scar with dome-shaped buds contrasts nicely with the semicircular leaf scars and plump buds on green ash. The larger, hairless leaves help distinguish these two in summer.

European Ash
Fraxinus excelsior

Though European ash has been planted in the Pacific Northwest for a long while, planting has really picked up in some towns since eastern cities stopped buying the popular cultivar 'Golden Desert' due to EAB, thus making for a cheap "developer tree," as some call it. You can find it mostly as a street tree.

SPECIES REMARKS: The dark black buds of European ash are the best clue in wintertime, especially if you can contrast them with the bright yellow twigs on popular cultivars. In summer (if there are no yellow twigs), look for leaves with more leaflets than other common species have and, in late summer, those dark black buds.

Large tree up to 80 feet (24 meters) tall but often much smaller, with rounded to upright canopy; most planted cultivars don't reach more than 30 feet (9 meters) tall. **BARK:** Like other ashes, gray with interweaving, crosshatched pattern. **TWIGS:** Though the normal species is gray to grayish-green, most cultivars have bright yellow twigs that stand out like swollen, vibrantly yellow sore thumbs; *buds distinctly black*; terminal bud pyramidal with rounded, plump accessory buds on either side. **LEAVES:** About 12 inches (30 centimeters) long with generally 9–11 nearly sessile leaflets that are coarsely serrated (other common species have fewer leaflets on average). **FLOWERS:** Unremarkable, but appear as dark reddish poofs in early spring. **FRUIT:** Cluster of paddle-shaped samaras, often with a flattened or notched (emarginate) tip and a slight twist.

193

Raywood Ash

Fraxinus angustifolia subsp. *oxycarpa* 'Raywood'
Syn. *Fraxinus oxycarpa* 'Raywood'

This tree (also called Mediterranean ash, narrowleaf ash) is a taxonomic menace, having only recently been fully included with narrowleaf ash as subspecies *oxycarpa*. Whatever its name, it's very common in the landscapes, especially west of the Cascades, where it's often found as a street or parking lot tree.

Medium tree up to 40 feet (12 meters) tall, with rounded, delicate-looking canopy, almost lacy; curiously often tipped and growing at an angle. **BARK:** Gray, crosshatched with interlacing ridges, creating a diamond-plate appearance. **TWIGS:** Rounded, green to gray, somewhat slender for an ash, with obvious lenticels; buds almost always *whorled in 3s*, especially near end of twig; terminal buds large, brown (not black); lateral buds brown, spherical. **LEAVES:** Up to 10 inches (25 centimeters) long with 7–9 narrow, widely spaced leaflets with serrations, making for a very feathery appearance; leaflets have hairs below only along base of midribs. **FLOWERS:** Small red poofs appearing in early spring; not very showy or helpful for identification. **FRUIT:** Mostly nonexistent, as it's a male clone; single samaras with wide wings and often a slight wave.

SPECIES REMARKS: If you find leaves and buds in threes, that's a sure sign it's Raywood ash: all others are in pairs. The cultivar 'Raywood' is almost certainly the only representative of this species in this region.

Flowering ash

Similar Species

Flowering ash (*Fraxinus ornus*) is usually found as a street tree and can be picked out quickly in midspring due to its large, showy clusters of white, strappy flowers on upright panicles. It also has smooth bark throughout its life, which is unique for this region's ashes (except some younger American white ashes).

It usually has seven to nine oval leaflets with acuminate tips, save for the terminal leaflet, which is slightly more rounded. The leaflet margins are serrated and undulating, and the midribs are pubescent near their base, setting this tree apart from American ash (which also has different buds; see table 11).

Blue ash (*Fraxinus quadrangulata*) and **Manchurian ash** (*Fraxinus mandshurica*) have been reported in tree inventories in the eastern portion of our region. But I'll level with you: they are quite rare in my experience. However, they are reportedly moderately resistant to EAB, so they may soon be the only ashes left standing (which will make identifying them that much easier, so there is that). Blue ash is notable for its four-angled twigs with small corky wings on each corner, and Manchurian ash has tufts of orange hairs where its leaflets attach to the main rachis (on the upper side of the leaf). These traits separate them from other common species, so if you see them, know you've got an interesting find.

Amur Corktree
Phellodendron amurense

Amur corktree feels new, but it has always been here. While it is being planted more often as a decent street and park tree, it still feels uncommon. Though reportedly invasive elsewhere, it hasn't shown bad habits in our region . . . yet.

Family: Rutaceae

Medium tree up to 40 feet (12 meters) tall, often spreading horizontally, with short trunk and thick limbs at maturity. **BARK:** Titularly corky in appearance; gray, hard, developing thick ridges that crisscross with old age. **TWIGS:** Stout, yellow to brown, glabrous with obvious lenticels; *base of leaf petiole completely covers bud*, so leaf scar nearly encircles bud or appears horseshoe shaped; buds nestled inside leaf scar, often bluntly pointed, covered in fine red hairs. **LEAVES:** Pinnately compound, up to 15 inches (38 centimeters) long with 7–13 dark green, oval to almost lanceolate leaflets, each with obvious light-colored veins, entire margins, acuminate tip; slightly pubescent mid-veins and woolly rachis. **FLOWERS:** Dioecious, appear in spring in large clusters; conspicuous, but not showy. **FRUIT:** Round drupe, starting yellow-green before turning black in fall.

SPECIES REMARKS: Japanese pagodatree and yellowwood also have pinnately compound leaves that cover the buds with their petioles; however, they are both alternately arranged, which immediately distinguishes them from Amur corktree. All other species and varieties listed as *Phellodendron* except for **Chinese corktree** (*P. chinense*) have been combined under *P. amurense*. Chinese corktree differs in that it has wider, longer leaflets on average that end with more abruptly pointed tips (mucronate) and have more rust-colored pubescence if any at all (see leaf on p. 184).

OPPOSITE, COMPOUND, PALMATE

The last of the oppositely arranged broadleaf trees are those with palmately compound leaves, and as it happens, just one genus fits the bill: *Aesculus*. The rare chaste tree (*Vitex agnus-castus*) also has oppositely arranged, palmately compound leaves, but contrary to its name, it does not often become tree sized in our region, so it isn't included here. There are also no significant trees with *alternately* arranged, palmately compound leaves in the Pacific Northwest.

Some trees that naturally have palmately lobed simple leaves have cultivars that appear palmately compound. Japanese maple is a quintessential example of

LEFT: *Yellow buckeye leaf* **RIGHT:** *Flowers and leaf of red horsechestnut*

this, as it has several popular "laceleaf" varieties whose lobes are all but separated by intensely deep sinuses. On one hand, this is a great illustration of why a compound leaf is still, in fact, just one leaf. On the other hand, it can derail a Stepwise Journey from the get-go. The good news is that later steps along the journey can help you check the accuracy of the preceding steps, by focusing on additional traits like the buds, flowers, or fruit.

The spiky husks of European horsechestnut (left) conceal the nutlike seeds (right), commonly found on the ground below the trees in fall.

HORSECHESTNUTS AND BUCKEYES

Aesculus

Horsechestnuts and buckeyes (family Sapindaceae) are the same type of tree, just with two different well-known common names. Essentially, if it's from Europe, it's a horsechestnut, and if it's from North America, it's a buckeye. Whatever their provenance, you'll know that you've got one because this is the only genus you're likely to see in the Pacific Northwest that has oppositely arranged, palmately compound leaves.

All species in *Aesculus* produce a "nut" in a leathery, often somewhat pokey husk. Botanically, these nuts are in fact giant seeds with a hard outer seed coat. The somewhat confusing fruit and the "leaves" of European horsechestnut (the most commonly found species in the region) superficially resemble sweet chestnuts (*Castanea* spp.), famous for being roasted over open fires. However, you'll always be able to quickly tell apart all the "chestnuts" by following the Stepwise Journey and starting with the buds, then defining the leaves. You'll notice that the "leaf" that looks like a sweet chestnut leaf is in fact a *leaflet* of the palmately compound horsechestnut leaf.

You can also look at the fruit to confirm, of course, because that's what really differentiates them, and they are substantially different. (Horsechestnuts are also poisonous, so don't go roasting them over an open fire.)

European Horsechestnut
Aesculus hippocastanum

Planted most often as a street or park tree, this is almost certainly the "chestnut" species you'll find growing in neighborhoods, especially older neighborhoods. It's developed a knack for seeding itself in cushy garden beds and moist natural areas when a parent is nearby, but it's certainly not invasive.

Large tree up to 75 feet (23 meters) tall, with upright, oval canopy; distinctive, large scaffold limbs. **BARK:** Dark gray, fissured, with circular, lifted plates that develop as it ages, very distinctive once you recognize it. **TWIGS:** Very stout, up to ¾ inch (2 centimeters) thick, with obvious large buds and leaf scars that nearly encircle twig; buds giant, conical, with obvious imbricate scales; lateral buds point out at 45-degree angles; buds sticky to the touch. **LEAVES:** Large, palmately compound with 7 obovate, serrated leaflets (very rarely 5), each 4–10 inches (10–25 centimeters) long; central leaflet is biggest, others smaller as you move left or right; each leaflet has abruptly acuminate tip and veins sunk into blade. **FLOWERS:** Borne on large, upright panicles in mid-spring, very showy; usually white with pink and yellow splotches and long, extended stamens. **FRUIT:** Round, shiny brown, nutlike seed covered in a green, regularly spiked husk borne in clusters in late fall; usually 1 seed per husk (rarely 2). **SPECIES REMARKS:** You'll start to notice this tree everywhere in larger cities. The leaves are the best identification characteristic, as they are palmately compound and opposite. However, the nut is just as good of a tell in the fall, and the large, sticky buds give it away in the winter, along with the unique bark.

WHEN IT'S NOT A EUROPEAN HORSECHESTNUT . . .

If you're walking down your street and find a tree with oppositely arranged, palmately compound leaves, nine times out of ten it's a European horsechestnut. The tenth time, it'll be a **red horsechestnut** (*Aesculus × carnea*), a hybrid between European horsechestnut and the confusingly named **red buckeye** (*A. pavia*). Red horsechestnut (see photos on p. 197) differs from European by being shorter and rounder, and almost always having *five ruffled leaflets* rather than seven and buds that are *not sticky*. In mid-spring, you'll see its ruffled magenta-red flowers like those of its red buckeye parent (again, that's *A. pavia*). Red buckeye—which has shiny, unruffled leaves with five leaflets and very small serrations—is very rarely found in the Pacific Northwest.

You may also run into a few other buckeyes, but they are equally as uncommon as red buckeye. **Yellow buckeye** (*A. flava*) is a large tree like European horsechestnut but has predominantly *five leaflets* that are smaller, more slender, and less coarsely serrated (see p. 197). It produces fruit with a *smooth husk* containing usually two seeds (see top right). **Ohio buckeye** (*A. glabra*) has *five unruffled, lightly serrated leaflets* too, but it's a shorter tree and produces fruit with an oblong husk and *many short prickles* (though not nearly as intense as European horsechestnut's); the fruit splits to release one to three seeds. **California buckeye** (*A. californica*) is a small, shrubby tree up to 25 feet (8 meters) tall, with five leaflets, *each with a long petiolule.*

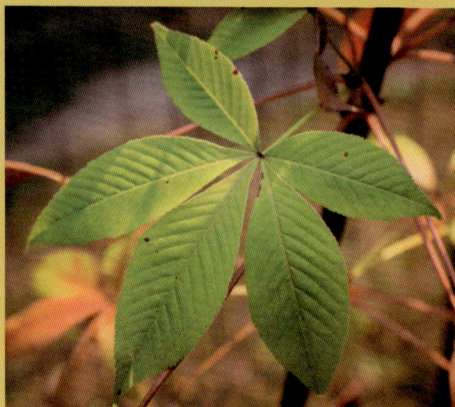

TOP: *Yellow buckeye fruit*
BOTTOM: *Ohio buckeye leaf*

ALTERNATE, SIMPLE

The majority of our species are found in this section, making them appear somewhat more complicated and a bit more intimidating. But remember that every tree has a unique combination of traits that will help you narrow down species by eliminating other options. By looking at the buds and leaves, you've already eliminated the oppositely arranged trees and those with some kind of compound leaves. By familiarizing yourself with the other attributes of the trees in this guide, the next steps become a bit easier.

With the sections organized by leaf type, identifying a tree by its leaves is as simple as getting to the correct section (i.e., alternately arranged trees versus oppositely arranged), then finding the leaves that most closely match what you have in hand and confirming specific details. But becoming familiar with the unique traits of a genus or species can help you use the Stepwise Journey more quickly. For example, most elms tend to have leaves with oblique bases, so if you see that trait, you know you have an elm. Likewise, if you see bands of horizontal lenticels on the bark, then the cherries and the birches are good places to start. The principles of the Stepwise Journey are the same, but it can go even faster if you take some time to learn which trees have which traits.

Also remember that I've included helpful lists of unique attributes at the end of the Stepwise Journey section. There you'll see the trees in this guide grouped by their distinctive traits, like fruit type, fruit color, bark attributes, and leaf colors. Just after these lists, I've provided a calendar of flowering times that outlines when different trees bloom across the seasons. These should help you get started, but if you flip through until you see a leaf that looks similar, you'll likely be close.

THE BIRCH FAMILY

The members of this family, Betulaceae, consistently get misidentified, as they are mistaken for one another more often than any other family, save for probably the rose family. It's worth taking a closer look at what differentiates them because, in the Pacific Northwest, all but one of the six genera that make up the family are well represented. The common trait of Betulaceae is the catkin, that wonderful flower/fruit type that looks more like the cone of a conifer than a flower—which has led to an awful lot of confusion for beginning botanists who hear the term "alder cone" when they learn about alders and question everything they know . . . or was that just me?

A **catkin** is considered an **inflorescence**, or a collection of flowers, bracts, and the stalks (pedicels and peduncles) they're attached to. In this family, they're essentially long, pendulous spikes densely packed with many unisexual flowers, each subtended by a scale-like bract. In early spring, both the staminate (male) and pistillate (female) catkins bloom, the former releasing their pollen, then quickly falling away. The pistillate catkin, however, once fertilized, transitions into an **infructescence**, or a collection of several individual fruits. For Betulaceae, these fruits are **nuts**, or rather **nutlets** in some cases, all densely packed together with their bracts in their catkins.

The characteristics of these nuts and nutlets and their clusters and catkins differentiate the genera of this family. Only the hazelnuts (*Corylus*) have done away with their female catkins altogether, reducing them to just a few flowers, each growing into a proper nut the size of an acorn. The rest can be differentiated by how their seed catkins look and act (see table 12).

The pollen catkins of Italian alder (Alnus cordata, left) and Turkish filbert (Corylus colurna, right) are very similar, firmly placing them within the same family.

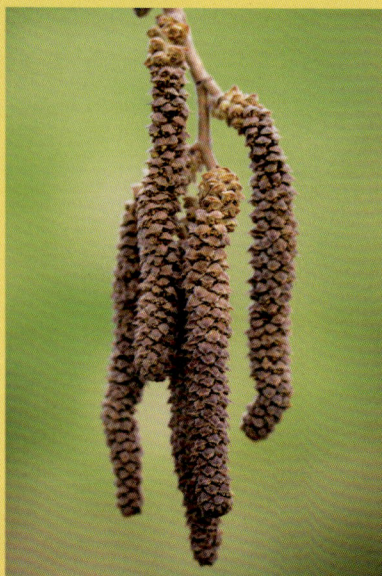

TABLE 12. QUICK GUIDE TO THE BIRCH FAMILY

GENUS & DESCRIPTION	LEAF	FRUIT
Birches (*Betula*) Leaves smaller than others, papery; bark has horizontal lenticels, often exfoliates; usually white, gray, or bronze; catkins fall apart when mature		
Alders (*Alnus*) Thick, doubly serrated leaves; buds set on short stalks (exception on shrubby native species); catkins woody, resemble cone of a conifer		
Hazelnuts (*Corylus*) Large, fuzzy, papery leaves with jagged serrations; often shrubby save for 1 rare species; fruit (hazelnut) with top covered by involucre		
Hornbeams (*Carpinus*) Leaves more lanceolate than others, singly or doubly serrated, with straight veins; bark smooth, gray; catkins large, with serrated, papery wings attached to smaller nutlets		
Hop-hornbeam (*Ostrya*) Leaves papery, doubly serrated, similar to hornbeams; bark flaky, exfoliating in vertical strips; catkins made up of inflated, pillow-like bracts with nutlets inside, like tiny almonds		

CLOCKWISE FROM TOP LEFT: *Himalayan white birch bark; river birch pollen catkins; upright seed catkins of yellow birch*

BIRCHES

Betula

Birches are set apart from others in their family by their dehiscent seed catkins, which fall apart at maturity, releasing their fluttering, winged nutlets to the wind. They also have the smallest leaves on average and the most flamboyant bark, the latter being the easiest way to pick one out of a crowd.

There are a few species native to the Pacific Northwest, but most are small shrubs. The one tree in the region (paper birch, which in fact is likely three different species, but more on that later) grows naturally only in the far northern and eastern areas, but it's planted quite often throughout, along with a bevy of ornamental species from Eurasia and eastern North America.

The best way to get to know your birches is to collect some leaves and see how they differ: each has slightly different characteristics, and they become quite clear when they are put next to one another (see table 13). The bark will also help separate them, but it can be a little trickier because of subtle changes in texture or color with age. With a little practice, though, you'll be amazed at how easily you'll be able to pick out their differences.

Bronze birch borer is a native North American insect that attacks all birch species by boring under the bark and feeding on the cambium below, often girdling the tree and causing branch dieback or death. Introduced species are highly susceptible, while native trees are less so, river birch being most resistant. It adds a unique macabre ID trait to our birches: if it's half-dead, you may lean toward a nonnative species over a native one.

TABLE 13. QUICK GUIDE TO BIRCHES

SPECIES & KEY TRAITS	LEAF	BARK
European white *B. pendula* Long, wispy, drooping twigs (often half-dead); dark furrows on lower trunk; leaves small, triangular, with big serrations; twigs with prominent lenticels, no hairs		
Downy *B. pubescens* Twigs not drooping, leaf base more rounded, bark grayer than European white's, new twigs are fuzzy		
Himalayan white *B. utilis* subsp. *jacquemontii* Upright (often half-dead) canopy; stems covered in smooth, white bark; leaves cupped, teardrop shaped; coarse single serrations		
Paper *B. papyrifera* White bark turns blemished gray with age, exfoliating horizontally; leaves narrow, doubly serrated, not cupped, sometimes wavy margin		
River *B. nigra* Leaves larger, more triangular, with intense double serrations; orange-bronze exfoliating bark		

SPECIES & KEY TRAITS	LEAF	BARK
Manchurian *B. pendula* subsp. *mandshurica*; syn. *B. platyphylla* Leaves like European white's, but distinctly triangular; bark not furrowed but grayish, with horizontal lines		
Monarch *B. maximowicziana* Oval leaves with minute serrations and distinctly cordate base; bark exfoliates wildly with age		
Yellow *B. alleghaniensis* Large leaves with sharp, uniform serrations; bark bronze-brown, exfoliating in small horizontal strips; catkins upright		

European White Birch
Betula pendula

You can find this tree almost everywhere west of the Cascades, from Anchorage to southern Oregon. Though planted in yards and along streets, it's nominally invasive and routinely found along roads and natural areas.

Medium tree up to 50 feet (15 meters) tall, with overall upright habit, but with *long, slender, pendulous branchlets*, making for a wispy texture. **BARK:** Light brown, warty when young, then bone-white; with age becomes rough, broken up by black furrows; *does not exfoliate like other birches.* **TWIGS:** Light brown, slender, smooth, *glabrous*, with warty lenticels; buds small, with imbricate scales, usually mottled in green and brown. **LEAVES:** Ovate to triangular, often slightly diamond shaped; 1–2 inches (2.5–5 centimeters) long, with small glands covering both surfaces; doubly serrated, with long tapering apex; a cutleaf variety is commonly found. **CATKINS:** Pollen cat-kins green to brown in winter, 1.5 inches (4 centimeters) long, held in clusters at end of twig all winter; elongate to double length in spring to release pollen; fruit catkin is 1–2 inches (2.5–5 centimeters) long, green maturing to brown in fall, disintegrating and releasing hundreds of tiny, winged nutlets.

SPECIES REMARKS: The white bark broken up by black furrows on the lower trunk quickly sets European white birch apart; all others remain smooth or exfoliate in many layers. If you're west of the Cascades in the US and see a wild birch that wasn't planted, it's likely this species (if you're farther north, see p. 210).

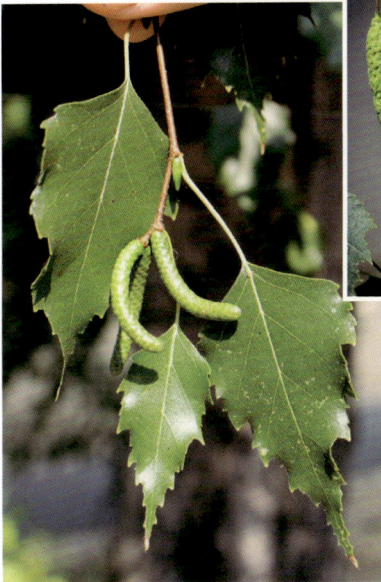

Himalayan White Birch

Betula utilis subsp. *jacquemontii*
Syn. *Betula jacquemontii*

This is the second most common species in the Pacific Northwest, often confused with paper birch due to its smooth, white bark. It's popularly sold in big-box stores, and you'll see it as a yard and street tree, but it's almost always half-dead due to drought stress and bronze birch borer.

Medium tree up to 50 feet (15 meters) tall, with upright, rounded canopy, often somewhat shaggy in appearance, but not drooping. **BARK:** Creamy white, smooth, with obvious horizontal lenticels; thin, exfoliating in horizontal strips with age; *maintains uniformly white, smooth bark through maturity.* **TWIGS:** Slender, light brown with pubescence on newest growth, but glabrous with age; obvious lenticels; buds sharply pointed, green turning brown. **LEAVES:** 2–3 inches (5–8 centimeters) long, ovate, teardrop shaped with wide, rounded base and acuminate tip; *leaf slightly cupped (not wavy),* uniformly serrated margins.

CATKINS: Similar to other birches; not wildly helpful for identification.

SPECIES REMARKS: The blemish-free bark at all ages gives this species away; no others are as uniformly white through middle and old age. The cupped leaves are the other great clue, unique to it alone. But honestly, just look for skinny, bone-white, half-dead trees struggling in planting strips and front yards: you can be pretty sure they are this poor thing.

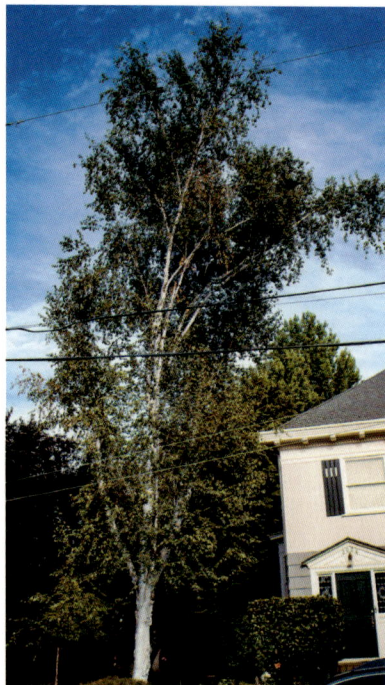

Paper Birch
Betula papyrifera

Though native to the northern half of our region, this tree is widely planted for its beautiful bark outside its natural range. In the north, it's everywhere, from natural areas to yards; in the south, you'll find it almost exclusively where it has been planted.

Large tree up to 80 feet (24 meters) tall, with upright habit, vase-shaped canopy, often somewhat irregularly shaped. **BARK:** White, smooth, with horizontal lenticels when young, fading to a darker gray with age, peeling in horizontal strips; becomes rougher with age but not furrowed. **TWIGS:** Slender, reddish-brown, slightly pubescent, with warty glands; buds green, turning brown through winter, bluntly pointed, with tiny hairs along margin of scales. **LEAVES:** 2–3 inches (5–8 centimeters) long, ovate, comparatively narrow with tapering tip, rounded or cuneate (wedge-shaped) base; margin irregularly doubly serrate (nearly dentate); flat to wavy (not cupped), with pubescence along underside of veins and petiole. **CATKINS:** Pendulous, similar to others.

SPECIES REMARKS: Paper birch is planted less frequently today in favor of Himalayan white birch, but it's still common. Separate it from Himalayan by its flat, narrower leaves and grayer bark on mature trees, and from European by its smooth (not furrowed) bark and bigger, more oval-shaped leaves.

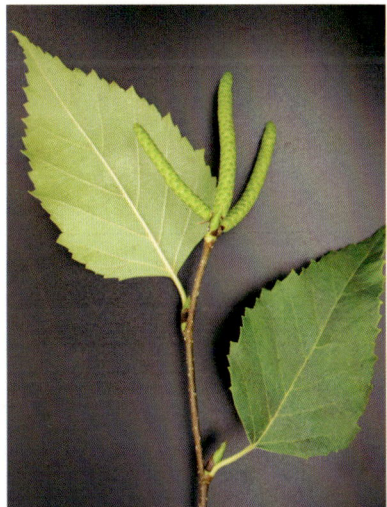

TWO LOOK-ALIKES IN THE NORTH

In the northern reaches of our region, where the wild forests and the urban or suburban landscape often intermix in terms of their tree species, two birch trees are causing some controversy: Alaska paper birch (*B. neoalaskana*, also called resin birch) and Kenai birch (*B. kenaica*). Both are variously listed as their own species, varieties of one or the other species, or as varieties or synonyms of European white birch and paper birch, respectively. Most recently they've been proposed to be separate from paper birch, and perhaps closer to a single species than previously thought.

While their genetic relationships are being worked out, I'm content to cover a few morphological differences for those of you in the far north who may have a birch species or two growing natively in your backyard (see table 14). A big point to note, however, is their ranges. You may only have one of these species growing in your area, sneaking into the landscape, which could make your life easier. (And remember, don't confuse them with your local shrubs either.)

TABLE 14. QUICK GUIDE TO THE WHITE BIRCHES OF THE NORTH

BARK	TWIG	LEAF SHAPE, MARGIN, AND BASE	LOCATION
Paper birch (*Betula papyrifera*) | **Tree Size:** 80 ft. (24 m)			
Creamy white, peeling, with pale lenticels	Slightly pubescent, becoming glabrous, sparsely covered in warty glands	**Shape:** Ovate, 2–3 in. (5–8 cm) long **Margin:** Doubly serrate to dentate **Base:** Rounded to wedge shaped	More upland, Southeast Alaska only, then south and east through north of region, barely into Washington and Idaho
Alaska paper birch (*Betula neoalaskana*) | **Tree Size:** 50 ft. (15 m)			
Reddish to off-white, peeling, with black lenticels	Densely covered in warty glands that conceal actual stem	**Shape:** Oval to deltoid, 1.5–2.5 in. (4–6.5 cm) long **Margin:** Coarsely doubly serrate, hairless **Base:** Broadly wedge shaped, with entire margin, longer petiole	Lowland, boggy soils in Alaska, far northeast British Columbia, northern Alberta
Kenai birch (*Betula kenaica*) | **Tree Size:** 40 ft. (12 m)			
Darker gray to white, with pink tinge and black lenticels	Pubescent, sparsely covered in glands	**Shape:** Deltoid, 1.5–2 in. (4–5 cm) long **Margin:** Doubly serrate, often with ciliate hairs **Base:** Rounded to flat	Uplands, rocky and Arctic areas in Alaska or Yukon only

River Birch
Betula nigra

Planted more often as a land-scape, park, and street tree due to its toughness and resistance to bronze birch borer, river birch is gaining in popularity by the second. It's native to eastern North America and is probably the most planted birch tree in new plantings.

Large tree up to 70 feet (21 meters) tall, with upright but spreading, open canopy, often as wide as tall; commonly planted with 3 main stems for apparently misguided aesthetic reasons. **BARK:** Smooth, *bronze to salmon colored* when young, but soon *exfoliating in raucous, ruffly strips and patches*, creating a striking mosaic of browns, reds, whites, pinks, grays; with age becomes rougher, ridged but maintains colors. **TWIGS:** Slender, reddish to copper, pubescent at first, then glabrous, with regularly spaced warty glands; buds pointed, chestnut brown, slightly appressed to twig. **LEAVES:** 2–3 inches (5–7.5 centi- meters) long, ovate to diamond shaped; lighter green than other birches, hairless below save for a few tufts along vein unions and a downy petiole; margin intensely doubly serrated, nearly lobed. **CATKINS:** Pollen catkin similar to other birches, but seed catkin sits upright and is much thicker, with larger, messier bracts.

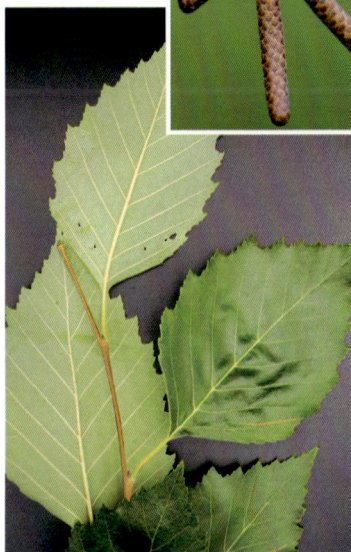

SPECIES REMARKS: If the bark doesn't immediately set this tree apart, then the leaves will. To me, they resemble a stylized drawing of a flame due to the much more prominent, lobe-like serrations as compared to all our other birch species.

LEFT: *The upright seed catkins of monarch birch* **RIGHT:** *Comparison of downy birch (left) and European white birch (right)*

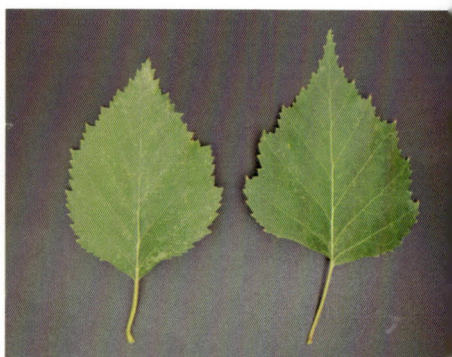

Similar Species

Downy birch (*B. pubescens*) is reported in the swamps and lagoons around Seattle, and it's often confused with European white birch, a close cousin. Despite their close relation, downy birch differs in having smooth yet pubescent twigs (rather than hairless yet warty), more rounded leaves with singly serrated margins, and dull grayish-white bark lacking intense black furrows (see table 13).

Manchurian birch, also known as Japanese birch or Asian white birch, is a tree by many names: historically it was *Betula platyphylla*, but it's now called *B. pendula* subsp. *mandshurica*. Regardless, this "species" is likely to be found planted in the colder eastern portions of our region where hardiness is of the utmost importance. Specifically, you're likely to see Dakota Pinnacle, a very upright selection (a good identification characteristic), planted as a street tree. Its leaves are smaller than most of our common species, around 2 inches (5 centimeters) long, strongly triangular with a flat, oblique base (the leaf blade is often offset on either side of the petiole); it holds its leaves late into the fall and breaks its buds weeks earlier than most other trees. The bark exfoliates similarly to Himalayan white birch but is more yellowish white.

Monarch birch (*B. maximowicziana*) has shaggy, exfoliating bark similar to river birch and its leaves are larger than our common birches, at 3–5.5 inches (7.5–14 centimeters) long, with only minute serrations and a telltale heart-shaped (cordate) base.

Yellow birch (*B. alleghaniensis*) also has bark like river birch, but it's darker red and exfoliates in small strips when young, becoming more yellowish gray and rougher with age. The leaves are oval, up to 4 inches (10 centimeters) long, with finely serrated margins (see table 13, p. 205). Yellow birch also has short, upright-growing seed catkins, like river birch.

CLOCKWISE FROM LEFT: *White alder leaf; red alder catkin; black or European alder fruit*

ALDERS

Alnus

Alders are common in Pacific Northwest forests, ranging in size from tall forest trees to alpine shrubs. With just over forty species worldwide, there is a comparatively small proportion represented as native species in this region. What's more, nonnative species are not really planted ornamentally either. So, what we have is really all we've got.

Alders are often called "weed trees" because of their knack for growing just about anywhere. They sprout up after disturbances like fire, logging, avalanches, and development and tend to grow in areas other trees prefer not to, like boggy lowlands, roadsides, and places with poor soils, like gravel lots and streamsides. It's my opinion, however, that this moniker is entirely slander because alders, with the help of a symbiotic relationship with bacteria called *Frankia*, add nitrogen to the soil and significantly improve its quality and structure. So, these so-called weed trees are actually more like ecological first responders.

Alders are unique among Betulaceae because of their woody catkins, which superficially resemble the cones of conifers. They also have relatively large, thick leaves and buds set on short stalks (except for Sitka alder).

Red Alder

Alnus rubra

By far the most common alder west of the Cascades, red alder quickly colonizes disturbed areas, especially where there is a decent amount of soil moisture. It grows happily in natural areas or otherwise ignored marginal landscapes.

(1.3 centimeters) long, look like pine cones set in small groups on short stalks.

SPECIES REMARKS: Red alder is the only one that has revolute leaves, which quickly and easily set it apart from other species. If no leaves are present, their smooth bark and large size will differentiate them, along with their angular twigs.

Medium tree up to 60 feet (18 meters) tall, usually with single stem, upright habit, airy canopy; pyramidal when young, becoming flat-topped with age. **BARK:** Greenish with prominent lenticels when young, becoming smooth, light gray with age; develops singular cracks; *covered in white lichen*, almost appearing white itself. **TWIGS:** Bronze to brown with lenticels, noticeably *angular in cross section*; buds red, *stalked*, pubescent, with just 2–3 scales. **LEAVES:** Oval, 3–5 inches (8–13 centimeters) long, thick, rough, tapering to dully pointed tip; veins prominent, straight, each ending in a large tooth along margin, itself serrated, *revolute* (meaning edges roll under). **CATKINS:** Pollen catkins long, 4–6 inches (10–15 centimeters) when flowering, releasing pollen in spring and falling away; set in late summer, *visible at ends of twigs through winter* (see photo, p. 213); seed catkins *woody*, ½ inch

Comparison of leaves of white alder (left) and red alder (right)

Similar Species

White alder (*Alnus rhombifolia*) is the most common species found in drier areas of the Pacific Northwest (west or east of the Cascades); it's also the only other native alder that reliably reaches tree size, though usually with multiple stems. It's not often planted, and its habitat does not overlap greatly with red alder, but it can find its way into town. Differentiate it from red alder by its *smaller, more rounded leaves* that *lack both a revolute margin and a pointed tip* (see photos, top left and on p. 213). Their twigs are not as angular, and their bark is only smooth when young, becoming rough and furrowed with age.

Italian alder (*A. cordata*) and **black or European alder** (*A. glutinosa*) may be found sparingly as landscape trees in the Pacific Northwest. Italian alder has smooth, heart-shaped leaves with cordate bases and tiny serrations along their margins, as well as larger seed catkins than other alders (see below). Black alder has thicker, paddle-shaped leaves with flat or slightly notched (emarginate) apexes and doubly serrated margins. You may also see cultivars of black alder with deeply cut leaves (see fruit on p. 213).

We also have two primarily shrubby, thicket-forming native species: **Sitka alder** (*A. alnobetula* subsp. *sinuata*) and **thinleaf alder** (*A. incana* subsp. *tenuifolia*). Both are part of circumpolar species complexes that include species from eastern North America, Europe, and Asia. You'll likely not see these in the built landscape, as they grow only in high-elevation forests or in the northern reaches of the region. If you see a bona fide tree, it's very likely not one of these.

LEFT TO RIGHT: *Italian alder leaf; Italian alder fruit; black alder leaf*

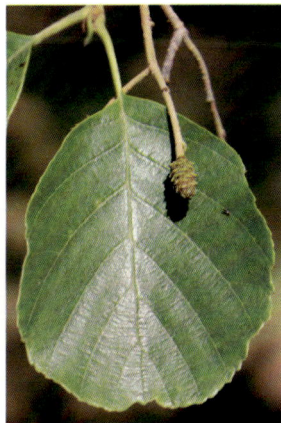

HAZELNUTS

Corylus

Three species of hazelnuts (called filberts 'round here, but also simply hazels) can be found in the landscape. Hazels have full-sized nuts instead of nutlets held on catkins, and each nut has an **involucre** around it, which is a bract-like appendage that looks like a sheath. The most common native species is at best a multistemmed shrub; the second most common is an orchard tree found mostly in Willamette Valley fields; and the only real tree of them all is quite rarely found, mostly as a street or park tree.

Beaked hazelnut (*Corylus cornuta* subsp. *californica*), our only native species, is found west of the Cascades and forms large, multistemmed, arching shrubs. It can get up to 15 feet (5 meters) tall, and its leaves are large, up to 4 inches (10 centimeters) long, and oval to rounded with coarsely serrated margins; they are papery and fuzzy on all sides. The nut is *tightly wrapped in an involucre* that forms a beak about 1 inch (2.5 centimeters) long at maturity (see table 12). Twigs are pubescent and zigzag at buds.

Common filbert (*Corylus avellana*), also known as European hazelnut, is the commercial species found in vast orchards in the Willamette Valley of western Oregon, and it's likely the producer of the nuts found in your bulk trail mix. It's often dispersed by animals and is thus frequently found growing in yards. Its leaves and form look just like our native species, but its nuts have shorter, fringed involucres that do not cover the whole nut, but rather look like lampshades around it.

Turkish hazel (*Corylus colurna*) is the only treelike hazel that we have, and it's a shame it's only found as a rare park or street tree. The upright, single-stemmed form will set it apart off the bat, but the leaves are also more acutely pointed, with more-pronounced cordate bases than the others. The nuts have involucres that look like wild Medusa heads at maturity. I often think its leaves resemble the dove tree's (*Davidia involucrata*), but their flowers and fruit are nothing alike.

CLOCKWISE FROM LEFT: *Beaked hazelnut leaf; Turkish hazel leaf; beaked hazelnut fruit; Turkish hazel fruit*

CLOCKWISE FROM TOP LEFT: *American hornbeam fruit; Japanese hornbeam leaves; hop-hornbeam leaf; American hornbeam leaf*

HORNBEAMS AND HOP-HORNBEAM

Carpinus **and** *Ostrya*

Hornbeams are relatively small trees, found in the Pacific Northwest primarily as street trees. Like all others in the birch family, they have pollen flowers arranged in long catkins. However, along with the hop-hornbeam, they differ from others in their family in that their winged nutlets are larger and arranged in loose clusters rather than tight catkins.

Over the last few decades, the Pacific Northwest's hornbeam repertoire has expanded beyond the popular and extremely common European species to include Japanese hornbeam, American hornbeam, and hop-hornbeam. They all stay relatively small (about 30 feet, or 9 meters, tall) and are planted mostly as street trees in smaller tree strips or below high-voltage power lines. All four can be confused with each other; their leaves help differentiate them (see photos, above and on p. 218), but often their fruit clusters are needed to be sure (see photos, pp. 218 and 219).

European Hornbeam
Carpinus betulus

Easily the most common of the hornbeams, this tree is planted as a street tree almost everywhere. It's not often found as a yard tree, and it doesn't grow wild in natural areas. You'll almost always find the upright (fastigiate) cultivar growing along roads, which is what is described here.

Medium tree up to 60 feet (18 meters) tall, pyramidal when young, but becoming very rounded, almost teardrop shaped with age; all limbs grow upward, making for a dense canopy lacking central leader. **BARK:** Smooth, gray, with very shallow, vertical etchings and often white lichen; stem not uniformly round but appears sinuous (muscly). **TWIGS:** Slender, greenish-bronze, covered in white lenticels; buds reddish-orange, large, pointed, incurved toward twig. **LEAVES:** Oblong to oval, up to 5 inches (13 centimeters) long, thin, papery with sharply doubly serrated margins and long, pointed tips. **CATKINS:** Pollen catkin short, releasing pollen in spring and falling away; seed catkin pendulous cluster of nutlets, each attached to 3-lobed bract; starts green in summer, matures to yellow-brown by fall.

SPECIES REMARKS: You can identify this species by its winter silhouette every time. The dense, upright growth creates a very characteristic, rounded canopy; no others really come close. It can be confused with European beech, especially when both are young, but hornbeams have serrated, duller-green leaves; smaller, less pointy buds; and 3-lobed bracts on their fruit.

LEFT TO RIGHT: *American hornbeam fruit; Japanese hornbeam fruit; hop-hornbeam fruit*

Similar Species

American hornbeam (*Carpinus caroliniana*) has leaves very similar to European's, only slightly wider (see photo, p. 217). Uniquely, its leaves turn orange-red in the fall, while the others remain a golden yellow. The fruits are small, pointed nutlets attached to large, three-lobed bracts with a few serrations along their edges. The nutlets are *borne in pairs* and are far more loosely clustered than others, and they have small, *bluntly tipped buds.*

Japanese hornbeam (*Carpinus japonica*) leaves are more lanceolate, a darker green, rougher, and *prominently ribbed* due to the lateral veins. They are sharply serrated and taper to a distinct acuminate tip (see p. 217). The clusters of nutlets are tightly layered together, each hidden by a bract that *lacks any distinct lobes* but is coarsely serrated. Buds are slender and pointed like European hornbeam's; do not curve inward as much.

Hop-hornbeam (*Ostrya virginiana*) has leaves like American and European hornbeams', but its fruit clusters look like Japanese hornbeam's. Rather than serrated bracts, though, it has layered, inflated sacs that look like paper almonds with seeds inside. Hop-hornbeam also contrasts with the others by having shaggy, light brown bark, not smooth and gray like the rest.

Hop-hornbeam bark

219

European Beech
Fagus sylvatica

This is the only species of beech commonly found in the Pacific Northwest. Botanically, beeches are closely related to oaks and chestnuts (see p. 330), though by appearance they look more like hornbeams. Usually quite large, this tree is found in yards, parks, and along streets. **Family: Fagaceae**

Large tree up to 75 feet (23 meters) tall, with wide, rounded, upright canopy; usually with low branches, splitting into multiple large scaffold limbs. **BARK:** Characteristically *smooth, gray*, only splitting with old age; often with white lichen; looks like giant elephant's leg; often carved into by wayward lovers. **TWIGS:** Slender, smooth, light brown to gray with sparse lenticels; *buds long, skinny, pointed*, up to 1 inch (2.5 centimeters) long; look like cigars. **LEAVES:** 2–4 inches (5–10 centimeters) long, oval; deep, lustrous green; margins entire, wavy (undulate), ciliate (fuzzy) when young; *5–9 straight parallel veins*; cutleaf, variegated, and purple varieties common. **FLOWERS:** Monoecious with separate male and female flowers on same tree; inconspicuous. **FRUIT:**

Smooth, brown, triangular nut, about ¾ inch (2 centimeters) long; usually 2 inside a 1-inch (2.5-centimeter) spiny husk that splits open in 4 parts.

SPECIES REMARKS: The pointy buds and spiky fruit are defining characteristics for beech trees, along with their bark. **American beech** (*F. grandifolia*) looks nearly identical save for its leaves, which are skinnier, gently taper to an acuminate tip, and have *nine to fourteen veins* ending with a serration along the margin (which mostly lacks cilia), but it's also almost never found in the Pacific Northwest.

Persian Parrotia
Parrotia persica

A very commonly planted street tree today, Persian parrotia (also called Persian ironwood) is fast becoming overplanted. It is small and tough and has pretty but subtle flowers and spectacular fall color, so it makes sense why we are seeing so much more of it. **Family: Hamamelidaceae**

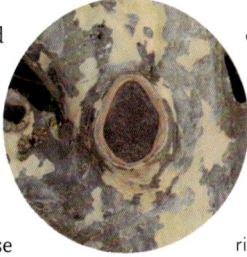

Medium tree up to 50 feet (15 meters) tall, upright growth, usually with many upright leaders, somewhat curvy branches. **BARK:** Smooth when young, but soon exfoliates in small scaly sections creating tan, white, and gray camouflage effect; stem undulates. **TWIGS:** Slender, gray to brown, slightly pubescent in spring; vegetative buds small, brown, tomentose; flower buds pea sized, spherical, brown, opening in mid- to late winter to reveal tiny red flowers. **LEAVES:** Up to 5 inches (13 centimeters) long, dark green, thick, ovate; veins angle down leaf, end in irregular serrations near the end of the leaf; margins undulate; mostly glabrous above; tend to have red tinge on new growth throughout summer. **FLOWERS:** Perfect but lack petals; bright red stamens are showy part, appearing as red puffs along twigs in early spring. **FRUIT:** Dry capsule, fairly unremarkable, similar to witchhazel.

SPECIES REMARKS: The bark and interestingly upright, rigid architecture set this tree apart in the wintertime; the curious flowers will give it away in spring; and in the summer, the leaves will be different enough from everything else but witchhazel to keep confusion at bay.

SPEAKING OF WITCHHAZELS

Parrotias are in the same family as witchhazels (Hamamelidaceae); witchhazels are far more widespread and well-known horticulturally, but far less treelike, so they earn only a brief mention here. Rarely reaching over 20 feet (6 meters) tall, they barely make it to tree size most of the time, usually growing up from a single stem that quickly splits into limbs that grow more outward than upward. The most commonly found species in the Pacific Northwest is **hybrid witchhazel** (*Hamamelis × intermedia*), usually just called witchhazel. It's a cross between Chinese witchhazel (*H. mollis*) and Japanese witchhazel (*H. japonica*).

Its branch architecture appears very angular, like an upside-down triangle balanced on the point. Its thick, wavy-margin leaves are broadly oval, up to 4 inches (10 centimeters) long, with veins that angle more down the leaf than out, similar to the parrotia's. Its flowers are spectacularly strange upon close inspection because of their four long, strappy petals that look like confetti stringers, ranging from yellow to red depending on the cultivar (and there are a lot of cultivars). It flowers in the dead of winter, from late January through February, making it easy to pick out of a crowd. If you see one flowering in fall, that's common witchhazel (*H. virginiana*), which ironically is rare to the point of being nonexistent in our region (though it is common in eastern North America).

THE WILLOW FAMILY

The willow family, Salicaceae, is a vast group that includes about 1,200 species across nearly five dozen genera distributed around the globe. At one time it included only the willows and poplars, cottonwoods and aspens, two mostly temperate genera (*Salix* and *Populus*, respectively) that fit quite nicely together due to their unisexual flowers in catkins and fluffy, cottony seeds.

Throughout the last half of the twentieth century, however, the family grew to include several genera that had previously been placed in a family that was described as a "taxonomic dumping ground": the Flacourtiaceae. Essentially, if botanists couldn't figure out where to classify certain species, they simply threw them in a catchall group, like an Island of Misfit Plants who didn't appear to belong anywhere. Luckily, we don't use these kinds of haphazard groupings anymore, and most of the family members got to stay together in Salicaceae, anyway.

Despite the new family additions, we here in the Pacific Northwest still have only the two original genera to deal with (the rest are mostly tropical species). The main differences for our purposes are in the form, leaves, and catkins. *Populus* species are large trees with generally large, wide leaves and pendulous catkins (both flowering and fruiting). *Salix* species are generally shorter trees or shrubs with smaller, more slender leaves and catkins that are more erect, growing up or out rather than simply hanging pendulously.

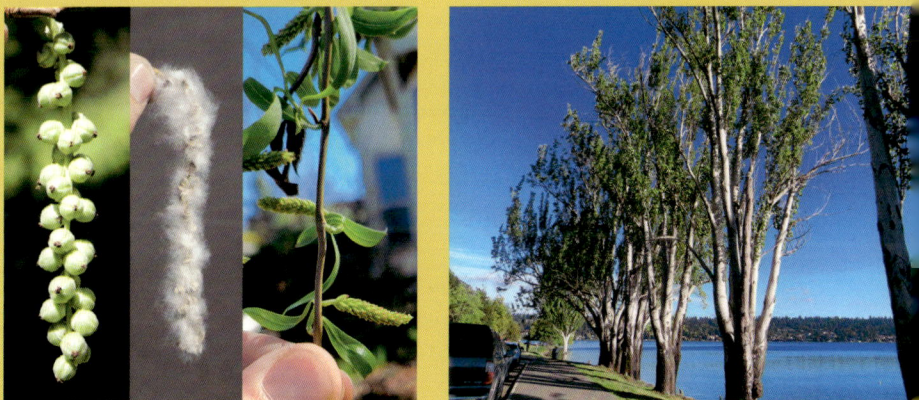

LEFT: *The pendulous fruit and cottony seeds of black cottonwood; the more erect catkins on a corkscrew willow* RIGHT: *Tall growing white poplars*

The bark and twig of Carolina poplar

POPLARS, COTTONWOODS, AND ASPENS

Populus

Poplar, cottonwood, and aspen are all different names for trees in a single genus, *Populus*. Very common in the Pacific Northwest, they can be found growing in wild forests just as often as in more urban landscapes. Some trees, like aspens, are preferred landscape trees—ill-advised in most circumstances, but alas, they're common in many newer landscapes both east and west of the Cascades. Others, like black cottonwood, are more often seen as unwelcome giants. All species in the genus are dioecious, with male trees producing pollen flowers on long catkins before the leaves appear, and female trees producing their infamous cottony seeds in pendulous clusters of small capsules in summer.

Like their close cousins the willows, cottonwoods and poplars are notorious for hybridizing willy-nilly (or should we say, willowy-nillowy). This trait, plus their flair for fluid leaf characteristics and their inclusion in breeding programs for fast-growing trees, leads to a complicated taxonomic situation.

In the Pacific Northwest, we have five bona fide species that are more or less native, even if they barely make it into our region naturally. Where their ranges overlap, the cottonwoods and poplars routinely form hybrids with each other (the aspens seem to not mingle as much with their more promiscuous cousins).

They also hybridize with the nonnative Lombardy poplar. Several of these hybrids are planted for cultivation, such as in large plantations east of the Cascades where you will find *Populus × generosa*, a hybrid between black and eastern cottonwoods, or in yards and parks where you will find Carolina poplar (*Populus × canadensis*), a hybrid between eastern cottonwood and Lombardy poplar. Unfortunately, that's just the tip of the iceberg, and wading through the taxonomy of all the species, subspecies, and hybrids is beyond the scope of this guide.

I've separated out the two most common species you're likely to find throughout the Pacific Northwest, which can be identified with little trouble: Lombardy poplar and black cottonwood. While these are really the only two species found west of the Cascades, all the other species and hybrids east of the Cascades join the mix the farther east you go, so I've included specific traits for each.

Black Cottonwood

Populus trichocarpa
Syn. *Populus tristis, P. balsamifera* subsp. *trichocarpa*

What a tree. Famous for getting really big really fast and then dropping huge limbs, black cottonwood is the most common tree of this genus you'll see in the western portion of our region, but it also grows commonly throughout. You'll find it in great numbers, mostly in or near lowland areas or along landscape margins.

Large tree up to 150 feet (45 meters) tall, with single, straight stem that splits about ⅔ of the way up into large, straight scaffold limbs that grow out at nearly 45-degree angles to main stem. **BARK:** Smooth, light gray when young, often with yellow lichen and black spots marking old branch locations; becomes thick, intense with age, developing large, light gray, flat-topped ridges, deep furrows. **TWIGS:** Light brown with few lenticels; buds large (up to ¾ inch, or 2 centimeters, long), green to brown, shiny, conical, sharply pointed; resinous (red colored, fragrant), smell of vanilla in spring. **LEAVES:** 3–5 inches (8–13 centimeters) long, ranging from heart shaped to triangular, with wedge-shaped or heart-shaped base; margin minutely serrated; upper surface green, underside pale silvery green with brown resin spots all over, appearing rusty; *petiole round.* **FLOWERS:** Dioecious; pollen flowers on male trees in long (up to 2 inches, or 5 centimeters), plump, reddish catkins similar to alders. **FRUIT:** Long catkin of practically spherical capsules, like a chain of pearls or the world's smallest, pendulous brussels sprouts; each capsule breaks open in 3 parts in late spring through early summer to release seeds covered in cottony filaments.

SPECIES REMARKS: One of the fastest and largest growing of our broadleaf trees, black cottonwood is often recognizable because of its perfectly straight stems and unique branch architecture—and because it's often far taller than the trees around it. The rounded petiole will set this species apart from others like eastern cottonwood and aspen.

Lombardy Poplar
Populus nigra var. *italica*

Historically planted as a wind-break tree all over the Pacific Northwest, and originally from Europe, Lombardy poplar is very common in towns and countrysides. Its upright profile makes it easy to spot, and often where there is one, there are many—not due to seeds (it actually doesn't produce seeds), but because of its tendency to sprout.

Large tree, often up to 90 feet (27 meters) tall, with very upright (fastigiate) habit and buttressed base; often single stemmed, but all limbs and twigs grow upright, so appears multistemmed. **BARK:** Gray, smooth when young, becoming rougher with age, developing long, curving ridges or slightly lifted plates; often with burls. **TWIGS:** Light gray, pliable, rather serpentine; buds similar to black cottonwood, only much smaller, held close to twig; hairless all around. **LEAVES:** 2–4 inches (5–10 centimeters) long, *triangular to diamond shaped*; finely crenate margins with *flattened petiole*, which makes leaves flutter in the wind. **FLOWERS:** Pollen catkins only, as this is a male clone.

SPECIES REMARKS: This is a fast-growing tree, and it's easily recognized by its extremely upright habit. It is also often half-dead due to a canker disease. Its form separates it from others even from afar.

Similar Species

Our other native cottonwoods and poplars are big trees, developing grand, capacious canopies held up on thick, outstretched scaffold limbs that arise from an often massive main stem. This is equally true for the hybrids they produce, even those with the intensely upright Lombardy poplar as a parent. They also develop intensely thick, furrowed bark with long ridges as they age, but when they're young, they have relatively smooth, gray bark. If you see a tree like this, especially east of the Cascades, you might have a poplar on your hands. The next step is to consider where it's growing, what its leaves look like, and what its buds look like. For more detailed treatments of the species and hybrids, check out Flora of North America online.

Balsam poplar (*P. balsamifera*) is closely related to black cottonwood. Some sources lump them together as varieties of balsam poplar, while others call them both synonyms of *P. tristis*, a Eurasian species. I'm dubious of this last lumping, and most sources still treat them all as separate species, so I'll maintain that convention here. Balsam poplar grows in the far north of the Pacific Northwest and overlaps with black cottonwood mostly in a few coastal areas of Alaska and northwest Canada, where they tend to hybridize. The primary difference between them is their fruit: instead of hairy, spherical capsules that split in three parts, balsam poplar's are more *pointed, smooth, and split in two*. Balsam poplar's leaves are also reportedly smaller, with a more rounded base, but it shares most other traits with black cottonwood.

Narrowleaf cottonwood (*P. angustifolia*) can be found near lowland drainages in the eastern portion of the region, especially in eastern Oregon, Idaho, and southwest Montana. It's similar to black cottonwood in most respects, but its *leaves are lanceolate with short petioles*, and its *fruit capsules are pointed, more*

TOP: *Narrowleaf cottonwood leaf*
BOTTOM: *Eastern cottonwood leaf*

egg shaped, and split open in two parts. This species is in the same subgenus as balsam poplar and black cottonwood, and it hybridizes with black cottonwood where their ranges overlap.

Eastern or plains cottonwood (*P. deltoides*) is the primary species in the eastern half of the continent; it barely finds its way over the Rocky Mountains and into the Pacific Northwest in a few places. However, it can hybridize with all the other species in this region, and

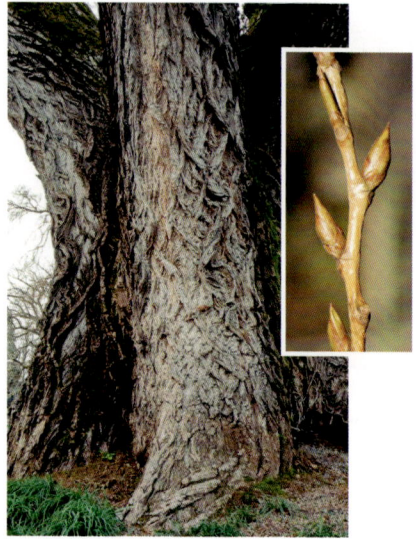

LEFT: *Eastern cottonwood bark and twig* **RIGHT:** *Carolina poplar bark and twig*

either it or one of its hybrids are the most commonly found cottonwoods in the landscape east of the Cascades; you'll often find it along the margins of the landscape or as a stately park tree. Unlike those of the previous species, its leaves have *flattened petioles* and usually *two to six glands where the petiole meets the leaf blade* (though one subspecies has none, as if it wasn't hard enough already). A few other marked differences are the large crenate teeth along the margin, which is also slightly translucent and fringed with tiny hairs (that is, it's ciliate); a lack of rusty splotches on the leaf underside; and the yellowish-green buds that exude a yellow resin, not red. **Carolina poplar**

Carolina poplar leaf

(*P. × canadensis*) is a commonly planted hybrid of eastern cottonwood and Lombardy poplar; it has no more than one gland where the petiole meets the blade, but otherwise it's similar to its eastern cottonwood parent.

Fremont cottonwood (*P. fremontii*) is native to the intermountain West and only barely reaches the southernmost extent of the Pacific Northwest natively, mostly east of the Cascades. You can find it sparingly along drainages and in lowlands near the edges of the landscape, where it takes advantage of the available water. It's closely related to eastern cottonwood and hybridizes with it where their ranges overlap. They both share several traits, such as flattened petioles; leaves with large, rounded teeth and translucent, ciliate margins; and buds that exude a yellowish resin. However, Fremont cottonwood differs in that it has no glands where the petiole meets the leaf blade, its buds are hairy, and its fruit is rounded rather than pointed. This is likely the least common of the cottonwoods in the region.

White Poplar
Populus alba

This invasive Eurasian species is common along highway margins and in ignored lowlands. It and its cultivars were once widely planted for their attractive foliage and bark but have since fallen out of fashion due to their invasive tendencies.

Large tree up to 70 feet (21 meters) tall but usually smaller; often with rounded, airy habit, but a few cultivars have more upright growth. **BARK:** Smooth, very white on younger stems, similar to aspen; with age, develops dark, diamond-shaped splits that ultimately develop into deep, crisscrossing, flat-topped ridges and furrows. **TWIGS:** *Covered with white hairs* that easily rub off; buds also densely hairy, reddish below hairs, plump, conical. **LEAVES:** Up to 3 inches (8 centimeters) long, variable from triangular or roughly egg-shaped (ovate), with coarsely crenate margins, to star-shaped, almost maplelike with 3 (sometimes 5) distinct lobes (especially on vigorous shoots); dark green on top, *underside densely covered with white, woolly hairs*; petiole round. **FLOWERS:** Dioecious; pollen flowers in yellow-brown catkins. **FRUIT:** A long catkin of *pointed capsules* filled with seeds covered in cottony fluff.

SPECIES REMARKS: The white pubescence on the twigs, buds, and leaf undersides, as well as the diamond markings on the medium-age bark, sets white poplar apart from similar species like the aspens and Lombardy poplar; no others have such intensely woolly parts.

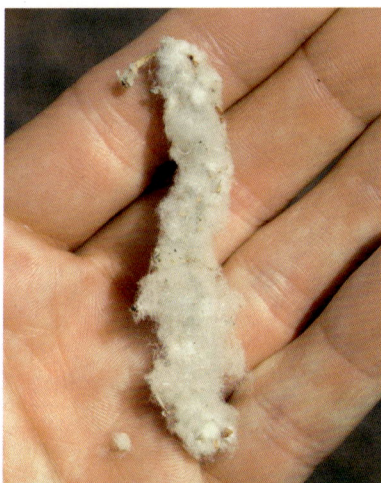

Quaking Aspen

Populus tremuloides

The most popular and romanticized tree of the mountains, quaking aspen is commonly found in our native forests and built landscapes. It tends to perform poorly on the west side of the Cascades, in my experience; but wherever it is, it always seems to do what aspens do, sending up shoots and then growing into a Hydra-like clump.

Medium tree up to 50 feet (15 meters) tall but usually smaller in the landscape; single stem, with skinny canopy, almost always taller than it is wide. **BARK:** White to light gray, visually striking, powdery; often unblemished save for dark spots where old branches once were; develops intensely dark furrows where damaged or on the oldest stems. **TWIGS:** Reddish-brown, hairless; buds small, shiny, conical, with pointed tips, darker brown than twigs, slightly incurved. **LEAVES:** No larger than 3 inches (8 centimeters) long, ovate to nearly round, with short, pointed apex and flat or slightly rounded base; margins finely crenate to bluntly serrate; glabrous (without hairs); *petiole flattened.* **FLOWERS:** Dioecious; pollen flowers in *fuzzy catkins* appearing in early spring, reminiscent of willows. **FRUIT:** Long catkin of pointed capsules filled with seeds covered in cottony fluff.

SPECIES REMARKS: The intensely white bark is the first clue to identifying quaking aspen, and its smaller buds, glabrous twigs, and pussy willow–like pollen flowers easily set it apart from other poplars. The rounded leaves flutter in the wind because of the flattened petiole (hence the name).

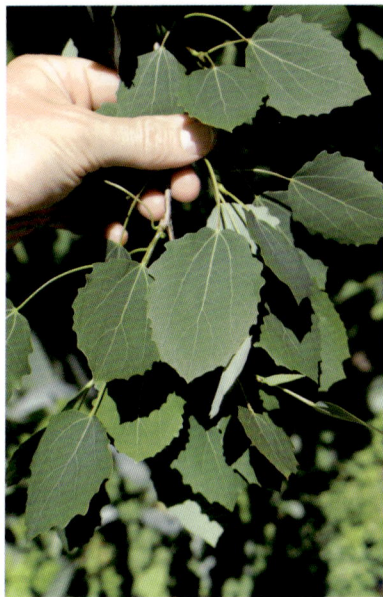

Similar Species

European aspen (*Populus tremula*—be careful with this one; that's just *tremula*, not *tremuloides*) is planted often west of Cascades north to Alaska. It looks very similar to quaking aspen. You'll nearly always find an upright cultivar called 'Erecta', whose branches grow straight toward the sky and whose twigs sort of flop out alongside. Its leaves are nearly the same size as quaking aspen's but are coarsely crenate along the margins and have a flat or cordate base. The bark is also grayer and more blemished, sometimes covered with yellow or black lichen.

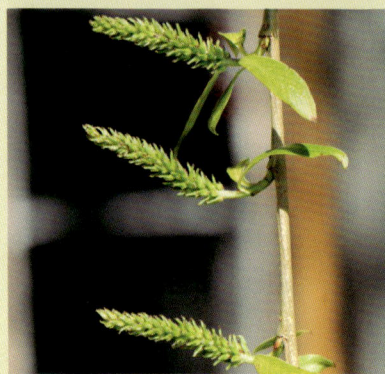

CLOCKWISE FROM TOP LEFT: *Weeping willow bark, leaves, and flower catkins*

WILLOWS

Salix

Willows upstage their cottonwood kin when it comes to confusing taxonomy. They tend to all look similar, they hybridize at will, and they have variable growth forms depending on where they're growing and how they got there. This, along with the name "willow"—used throughout history to refer to almost any tree growing next to a stream or with very long, thin leaves—is a recipe for chaos.

Luckily, you can gently float above the mire and focus your attention on only those few willows that tolerate the domestic environments of the Pacific Northwest. A few native species are found in cities and towns here, but more often you'll find a nonnative. Most of the native species are shrubs anyway, so if you see a larger tree, it's likely (though not necessarily) a nonnative species.

If you would like to really test your botanical acumen and explore the wild world of willows in the Northwest, the guides listed in the resources will offer some help. May nature have mercy on you.

Scouler's Willow
Salix scouleriana

One of the Pacific Northwest's more common native species, this willow can often be found in yards and more urban natural areas. It prefers wet habitats but does well enough in a variety of conditions. It's more treelike west of the Cascades but still tends to send up many stems and sprouts.

Small tree, usually not more than 25 feet (8 meters) tall, but in the right conditions can get much taller, with upright, rounded canopy, often with several leaders. **BARK:** Gray, mostly smooth when young, with diamond-shaped marks; splits into wide, flat ridges broken up by furrows with age, revealing orange in the cracks. **TWIGS:** Fairly stout, olive green to brown, with slight pubescence; vegetative buds red, pubescent, conical, pointed but flattened on the side against the stem; flower buds much larger, rounder, containing white, densely furry, silver flower catkins. **LEAVES:** Usually less than 4 inches (10 centimeters) long, obovate or oblanceolate, with rounded tips and entire, but sometimes wavy, margins; top dull green, hairless; underside lighter green with pubescence. **FLOWERS:** Dioecious; both catkins appearing in early spring; male trees have short catkins, less than 2 inches (5 centimeters) long, that start densely fuzzy, like a kitten's paw, expanding to send out tiny yellow flowers; female trees have similar catkins with red pistils. **FRUIT:** Hairy catkin with many small, woody capsules that release seeds covered in cottony hairs.

SPECIES REMARKS: Look for the obovate leaves of Scouler's willow to set it apart from other treelike willows in the area, like the Pacific willow (see p. 237). It's not uncommon to find Scouler's willow away from water too, which is a good clue, as other native willows prefer wet habitats.

WEEPING WILLOWS—WHERE TO START?

Most people are very familiar with weeping willow; it's a classic landscape tree that is very recognizable even from afar. However, the taxonomy of the species is far less familiar: the scientific names and classifications are moving targets, as species are combined and hybrids are declared and dissolved seemingly constantly. So, I'm going to break it down for you as best I can in terms of functional tree identification. Truthfully, I wouldn't blame anyone if they identified a tree as a willow and then stopped there. Willows are quite the handful.

Historically, the two primary species of weeping willow have been the golden weeping willow (*Salix alba* 'Tristis') and the weeping willow (*Salix babylonica* 'Babylon'). However, the weeping variety of *S. babylonica* is not exceptionally hardy and therefore not commonly found in the nursery trade these days, so it's unlikely you'll find it in the Pacific Northwest. What has been historically called *S. alba* 'Tristis' is today regarded as a hybrid of *S. alba* and *S. babylonica* that has been given at least two names: *S. × salamonii* and *S. × sepulcralis*, with the variety name *chrysocoma* added on. These hybrids are themselves in dispute, as they are now recognized as synonyms of *S. × pendulina*, which is a three-way hybrid with *S. alba*, *S. babylonica*, and crack willow (*S. × fragilis*), which itself is a hybrid of *S. alba* and *S. euxina*, another European species. You can see how this quickly becomes exhausting.

All things considered, most authorities hold that the most common weeping willow in the Pacific Northwest is whatever used to be called *Salix alba* 'Tristis', which as of this writing I believe is *S. × pendulina* var. *tristis*. If you run across a resource that says any of the other names, it is probably safe to assume we're all talking about the same weeping willow tree.

Golden Weeping Willow

Salix × pendulina var. tristis
Syn. *Salix alba* 'Tristis' and others

Golden weeping willow (also just weeping willow) is the tree you've come to recognize in gardens and parks growing pendulously and gracefully near streams and wetlands. The hardiest weeping willow, it's a hybrid between European and East Asian species.

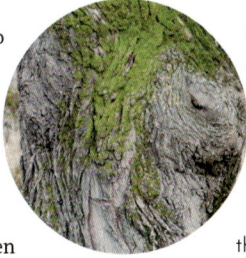

Large tree up to 60 feet (18 meters) tall but variable in height, with large, rounded, extremely pendulous crown; stems often large, contorted with wide-spreading habit. **BARK:** Gray to yellow-brown, thick, furrowed, often spiraling around stems. **TWIGS:** Long, slender, very pendulous, often to the ground; true species green to brownish but most often *bright-yellow-stemmed selections found*; pubescent, especially when young; buds small, yellow, appressed to stem. **LEAVES:** Up to 6 inches (15 centimeters) long, lanceolate, with serrulate margins; pale green, glossy above, with some silky hairs when first emerging but soon hairless; paler green to whitish below; *lack glands at base of petiole*; generally no persistent stipules. **FLOWERS:** Dioecious, both catkins appearing in early spring; male trees have catkins up to 2 inches (5 centimeters) long, with many tiny, somewhat showy yellow flowers; female trees' catkins slightly smaller, with green, less-showy flowers. **FRUIT:** Mature female catkin with many green, 2-valved capsules that split open, exuding many tiny seeds covered in silky hairs.

SPECIES REMARKS: As noted above, the traits of this tree can be variable, but suffice it to say that if you see an intensely weeping willow tree, it's probably this one or some cultivar of it.

235

Corkscrew Willow
Salix babylonica var. *pekinensis* 'Tortuosa'
Syn. *Salix matsudana*

A common ornamental willow west of the Cascades, corkscrew willow (also called curly-leaf willow) is another victim of recent taxonomic upheavals, as it's been subsumed into *S. babylonica*. You will mostly find it growing as a yard tree.

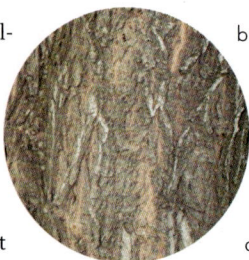

Medium tree up to just over 30 feet (9 meters) tall, with upright habit, very contorted stems, wrinkly architecture; short-lived, often with broken or dead limbs. **BARK:** Yellowish-brown to gray, with flat-topped ridges separated by shallow fissures when young; interweaving ridges and deeper furrows with age. **TWIGS:** Distinctly curly or contorted, slender, greenish-yellow; buds small, appressed to twigs, yellow to beige. **LEAVES:** Up to 4 inches (10 centimeters) long, lanceolate, wavy or recurved especially near tip; margins serrated; very distinctive. **FLOWERS:** Dioecious, but curly-stemmed varieties found in the landscape are female clones; yellowish-green catkins appear in spring (see p. 223). **FRUIT:** Not often present, if at all; similar to weeping. **SPECIES REMARKS:** The form and branch architecture of corkscrew willow should be the first clues to its identity; no others look similar except for a very small, curly shrub variety of the common filbert, but its catkins and leaves are starkly different.

Similar Species

Pacific willow (*S. lasiandra* var. *lasiandra*) is often found growing in drainages or near wetlands across the region, both in built environments and in more wild areas. It gets to be tree sized, but it also tends to break apart and regrow, so it often looks like a clump of vegetation. Its leaves are lanceolate, 2–7 inches (5–18 centimeters) long, glossy green on top and glaucous white below (the other variety, *caudata*, which tends to be found east of the Cascades only, is not glaucous below). The petioles have *usually two glands where they meet the blade* and often a stipule too. Importantly, when you bend the twigs, they *do not snap cleanly off* the stems, but bend and otherwise tear; the nonnative crack willow (see following page), which looks similar, will audibly snap.

White willow (*S. alba*) is an upright, tall willow found throughout the region. It looks like a non-weeping version of a weeping willow. The leaves are pubescent all season long and usually have two dot-like glands on the petiole near where it meets the blade. Its twigs are a bright orange-yellow. In the landscape, you're likely to see either weeping willow, corkscrew willow, or this white willow, and luckily they are simple to tell apart by form alone: weeping, corkscrewy, or upright.

ABOVE: *Pacific willow* BELOW: *White willow*

237

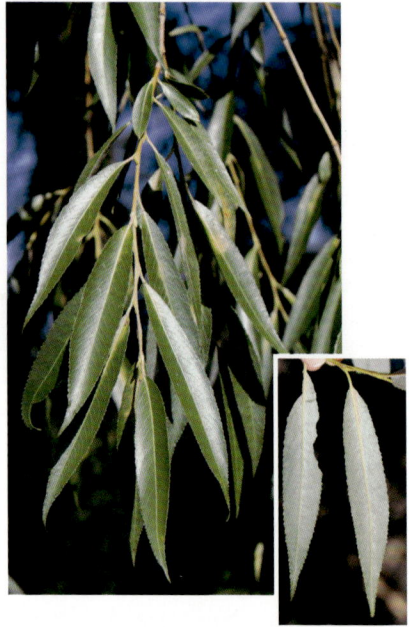

Crack willow (*S. × fragilis*) has been planted extensively in the Pacific Northwest, as well as across the continent, and is considered an invasive species. It's an upright to spreading tree growing to about 60 feet (18 meters) tall, with a round canopy, often filled with large, broken limbs. Its leaves are strongly lanceolate, green on top and whitish below, with two glands on the petiole where it meets the leaf blade, and no persistent stipules. Importantly, its *twigs snap when bent, making a distinct cracking sound* (hence the name). The weeping willow will also do this, but its form will set it apart. You'll find crack willows near wetlands, drainages, and remnant windbreak plantings.

Peachleaf willow (*S. amygdaloides*) is native to a few areas in the far east of the region, with most of its native range east of the Rockies. It can find its way into lowland areas in its range as cottonwoods do, so it deserves a mention. Peachleaf willow has broader leaves, often double the others in width, and they're glaucous below with no stipules or glands (almost always; newer shoot may have stipules, as in the photo to the right). Their stems will bend and tear, not snap.

TOP ROW: *Crack willow*
ABOVE: *Peachleaf willow*

Common Hackberry
Celtis occidentalis

This is the only species of hackberry commonly found in the landscapes of the Pacific Northwest, usually as a park or street tree and often in bioswales because of its tolerance of poor, lowland soil conditions. **Family: Cannabaceae**

Large tree up to 60 feet (18 meters) tall, with rounded overall shape; tends to grow more upright when near other trees. **BARK:** Dark gray, *uniquely warty*; develops corky bumps, ridges at a relatively young age, becoming platy with age. **TWIGS:** Brown, smooth, but sometimes with slight pubescence, prominent lenticels; buds small, triangular, lighter gray than twigs; incurved, sitting on pronounced ledges, giving them a slightly zigzag shape; terminal bud grows nearly sideways at end of twig. **LEAVES:** 2.5–3.5 inches (6–9 centimeters) long, broadly triangular, but with pronounced *asymmetrical bases*

and long, tapering tips; *rough to the touch*, with serrated margins; 2-ranked along the twigs, often afflicted by galls. **FLOWERS:** Singular, inconspicuous, located below leaves and twigs on long stalks. **FRUIT:** Dark red, fleshy drupe, size and shape of a marble, borne below leaves, down twigs.

SPECIES REMARKS: The rough leaves with asymmetrical bases are a great clue for common hackberry in summer, along with the singular, round fruits beneath the twigs. Confirm it with the bark all year round; no other trees have warty main stems like this. **Hardy rubber tree** (*Eucommia ulmoides*) is an uncommon tree that looks like the hackberry, but its fruit is papery, and its leaves are darker green and wrinkly (rugose) on top; when ripped, they have strings of latex between the ripped halves.

BOTTOM RIGHT: *Hardy rubber tree leaf*

239

Japanese Zelkova
Zelkova serrata

A common street tree west of the Cascades that is becoming more common every day, Japanese zelkova can be a striking tree when left to grow large. But all too often varieties are planted that I fear will prove short-lived due to poor form and structure. Time will tell. **Family: Ulmaceae**

Medium tree, usually up to 60 feet (18 meters) tall, with vase-shaped to upright canopy; a lot of variation depending on cultivar. **BARK:** Gray when young, speckled with orange lenticels; with age, starts to exfoliate in rounded scales, revealing orange patches; lenticels present on and below scales. **TWIGS:** Slender, light brown, zigzag at each node; buds very small, red, conical; sometimes 2 buds per node: one for leaves, the other for flowers. **LEAVES:** Up to 2 inches (5 centimeters) long, narrowly ovate, dark green with several straight veins ending in bulbous serrations that curve toward leaf apex; *base symmetrical*, surface rough. **FLOWERS:** Fairly inconspicuous, appear at nodes; both male and female flowers on same tree. **FRUIT:** Small, nutlike drupe in leaf axils, often also inconspicuous.

SPECIES REMARKS: Zelkovas look very similar to a few species of elms, notably Siberian elm and lacebark elm. Its bark helps set it apart from Siberian elm, which has thicker bark, and its larger leaves with large serrations set it apart from lacebark elm, which has similar bark. Also look for its nutlike fruit; elms have papery samaras.

LEFT TO RIGHT: *A massive wych elm (*Ulmus glabra*) successfully identified; The upright form of English elm is typical of most species aside from American and lacebark elms.*

ELMS

Ulmus

Elms (family Ulmaceae), and specifically the American elm, used to be the most planted street and park trees—not only in the Pacific Northwest but also across North America. Dutch elm disease (named after the nationality of the scientists who found it) has killed off most of them elsewhere on the continent but has been slower to completely decimate those susceptible to it here. Nonetheless, those that are susceptible are still dropping like flies.

Elms can be challenging to identify for a few reasons. First, they all tend to look similar, so it takes a keen eye to really determine the most likely species (see table 15). Second, they've been bred so much to combat Dutch elm disease that what's left is a huge number of cultivars that show intermediate traits between their parents, who themselves were likely the product of intensely interbred stock.

When it comes to telling species apart, elms are a lot like catalpas: their differences can be subtle and dependent on the season. I have done my best to give a simple breakdown of the species and their traits in table 15. Much of this information was originally compiled and shared with me by Phyllis Reynolds, who wrote *Trees of Greater Portland*, which I highly recommend if you want to get better at elm identification (and find some of the most spectacular trees in the city).

TABLE 15. QUICK GUIDE TO ELMS

SPECIES & KEY TRAITS	FRUIT	LEAF
Lacebark *U. parvifolia* Leaves small (2.5 in, or 6 cm, long), dark green with equal base; samaras found in fall; bark orange, exfoliates in patches		
Siberian elm *U. pumila* Leaves skinny, symmetrical, with mostly even base; buds rounded; upright form; gray bark; circular samaras		
American elm *U. americana* Leaves up to 5 in. (13 cm) long, oblique base, papery, sharply doubly serrated; samaras fringed with hairs; vase-shaped form with interweaving bark		
European white *U. laevis* Leaves up to 4 in. (10 cm) long, very oblique, hairy below; buds sharply pointed; samaras fringed with hairs; form upright; bark gray, scaly; base fluted		

SPECIES & KEY TRAITS	FRUIT	LEAF
Dutch elm *U. × hollandica* Leaves up to 7 in. (18 cm) long, abruptly oblique, papery, wider than American's; buds rounded; samaras hairless; upright habit; gray scaly bark		
English elm *U. minor* Leaves 3.5 in. (9 cm) long, oblique base, smooth above, pubescent veins below; corky wings on twig; upright canopy; gray, vertically scaled bark		
Wych elm *U. glabra* Leaves up to 6 in. (15 cm) long, widest of elms, often with 3 large points, pubescent; usually extremely drooping cultivar, rarely large, upright tree; samaras oval		
Japanese elm *U. davidiana* Leaves up to 4 in. (10 cm) long, obovate, dark green, asymmetrical, glossy; ridged bark like American's; oval, wavy samaras		

Lacebark Elm
Ulmus parvifolia

This species has become more popular recently due to its medium size, attractive bark, and resistance to Dutch elm disease. You'll find it most often as a street tree, but it's not uncommon in parks and landscapes, and it does not seed itself in.

SPECIES REMARKS: The bark, leaves, and fruiting time are quick identifiers for this species. It can be confused with Siberian elm or zelkova, both of which have similar leaves. Siberian elm gets big and has rough, furrowed bark, and zelkova's leaves are larger, with more sharply serrated margins.

Medium tree up to 50 feet (15 meters) tall, with rounded canopy, delicate texture (for an elm). **BARK:** Distinctively scaly, gray and orange, covered with small orange lenticels; very similar to zelkova. **TWIGS:** Gray, slightly pubescent, zigzag in appearance; buds small, dark brown, minutely pointed. **LEAVES:** Smallest of regional elms, up to 2.5 inches (6 centimeters) long, with *nearly symmetrical base* and blade; dark green top, thick but not wide, brittle if bent; margin technically serrated, but appears more like notches. **FLOWERS:** Perfect, appear in mid- to late summer (not spring like others). **FRUIT:** Round to oval samara with seed in middle, small notch at the tip; *appears in fall* and may persist through winter; no hairs on edges.

Siberian Elm
Ulmus pumila

This is the Pacific Northwest's main invasive elm, growing wherever it can get a hold. Its drought and cold tolerance make it well adapted to interior habitats, and you can almost be sure that if you spot a scrubby elm popping up from a crack in the pavement, it's this species.

Large tree up to 70 feet (21 meters) tall, with rounded but upright canopy, often with many long, thin, dangling twigs, messy appearance; not uncommon to see broken limbs. **BARK:** Thick, light gray, similar to cottonwood; not shaggy. **TWIGS:** Slender, greenish to gray with slight zigzag appearance; buds *small, dark gray, globe-like*. **LEAVES:** 1–3 inches (2.5–8 centimeters) long, *nearly symmetrical at base*; similar to lacebark elm, only wider and longer; essentially *singly serrated*. **FLOWERS:** Perfect, appear before leaves in early spring, incon-spicuous. **FRUIT:** Small, nearly perfectly round samara with seed in middle and either *closed notch* at tip or none at all; no hairs around edge.

SPECIES REMARKS: Odds are if you're east of the Cascades and you see an elm growing where it doesn't look like it was planted, it's a Siberian elm. The symmetrical leaves and very round buds are excellent clues for this species, as well as the bark and upright, often skinny, messy form.

American Elm
Ulmus americana

American elm is a common street and park tree still found in good numbers in the Pacific Northwest, especially in larger towns. This is not an invasive elm, so you won't find it seeding itself in.

SPECIES REMARKS: The dark, interweaving, ridged bark, the crisscrossing branch architecture, and the seed with a halo of little hairs will set this species apart. Dutch elm has larger leaves that are more offset at the base, no hairs around the samaras, and lighter, vertically fissured bark.

Large tree up to 80 feet (24 meters) tall, with single stem that arcs upward and outward, creating distinctive vase-shaped canopy; large scaffold limbs crisscross, creating unique architecture. **BARK:** *Dark gray, deeply ridged, furrowed* on older trees; ridges interweave. **TWIGS:** Brown, moderately pubescent, often with zigzag appearance; buds *distinctly pointed*, angled away from twig. **LEAVES:** Usually 3–5 inches (8–13 centimeters) long, ovate with rough, papery texture; base usually offset, but not as dramatically as other species; doubly serrated margins, pronounced veins. **FLOWERS:** Perfect, appear in clusters in early spring, not wildly useful for identification. **FRUIT:** Flattened, circular samara with single seed in middle and shallow notch at tip; *fringed with hairs*; the size of a dime or smaller.

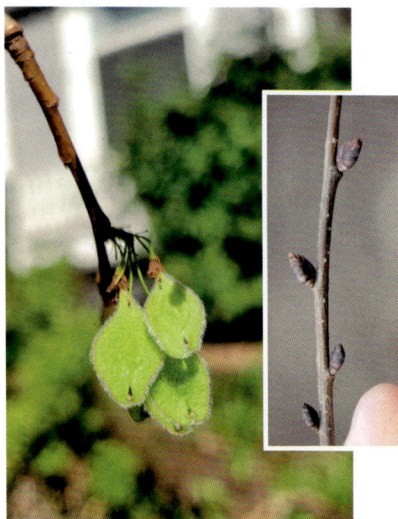

Japanese Elm
Ulmus davidiana

This is where the elms really go off the deep end. This species name is the result of a complex taxonomic reordering that combined *U. propinqua*, *U. japonica*, *U. wilsoniana*, and *U. davidiana* under one species name with two varieties. To add to the confusion, you're not likely to see the true species form planted, but rather one of several cultivars that were the result of hybridization and crossbreeding.

Medium tree up to 60 feet (18 meters) tall, with shaggy top of unruly foliage when young, but rounded, vase-shaped canopy with age; similar to American elm, but smaller. **BARK:** Like American elm, dark gray with interweaving ridges even when young. **TWIGS:** Light brown, glabrous, very zigzag; buds small, dark brown to blackish, conical with blunt point. **LEAVES:** Up to 4 inches (10 centimeters) long, *obovate* with very asymmetrical base; glossy green on top, often rough to the touch, with pronounced veins below; rigid, not papery, with very slightly doubly serrated margins and abruptly acuminate tip. **FLOWERS:** Perfect, appear in early spring before leaves. **FRUIT:** Oval, crumpled samara with small notch, often in large clusters.

SPECIES REMARKS: Cultivars like 'Accolade' and 'Emerald Sunshine' are popular for their form and resistance to Dutch elm disease, but there are plenty of other options out there. Between this species and the last three, your hands will be full figuring out which is which!

LEFT: *'Emerald Sunshine' leaf* **RIGHT:** *'Accolade' leaf*

247

Similar Species

European white elm (*Ulmus laevis*) is a relatively uncommon species in our region but it's easily confused with American elm so it's worth mentioning the differences. Its leaves are pubescent below and its buds even more pointed. Its bark is light gray and scaly, its base is notably fluted, and its habit is more upright.

Dutch elm (*Ulmus × hollandica*) is a hybrid between *U. minor* and *U. glabra*. Its leaves are bigger than American's, and its samaras are the size of a nickel and lack hair. The buds look like small ovals that have been slightly squished; they aren't pointed. The bark is light gray and rough, broken up irregularly into shallow, flaky ridges. It is also a very upright tree, not vase-shaped like American elm.

English elm (*Ulmus minor* subsp. *minor*, previously *Ulmus procera*) has similar bark and seeds to Dutch elm's, but smaller leaves with *intensely oblique bases*; they are rigid, smooth above, with prominent, pubescent veins below, all in a compact arrangement. Its buds are more conical than Dutch elm's, and it tends to develop corky wings on its twigs more often. This species is very common in the eastern portion of our region as a street tree.

Wych elm (*Ulmus glabra*) is almost never found as a large tree but rather as the strange, contorted mophead of a tree, Camperdown elm (*Ulmus glabra* 'Camperdownii'). Regardless of the form, the leaves are rounded, oblong, wider than most other species, and distinctly sandpapery to the touch. They tend to develop one or two pointed lobes on either side of the apex, making them look a lot like a hazel's. Their twigs are pubescent, with large, dark brown, conical, pubescent buds, and their fruit is oblong or roughly diamond shaped.

CLOCKWISE FROM TOP LEFT: *Classic linden fruit; littleleaf linden form; classic linden flower*

LINDENS

Tilia

Lindens (also called limes in European and basswoods in Eastern North America, family Malvaceae) are very common street trees in the Pacific Northwest, but they still sort of fly under the radar. In Europe, they are well-known as park and garden trees (specifically *beer* gardens in Germany), but for whatever reason they don't seem to have entered the common vernacular around here like maple or oak. Nonetheless, if you're in a town west of the Cascades, it's all too likely that there's one growing nearby.

All lindens have heart-shaped leaves and very curious flower/fruit clusters that quickly and easily set the genus apart. Appearing in early summer, their small, fragrant flowers are in pendulous **cymes** (flat clusters where the middle or terminal flower blooms first) attached to an unmistakable, paddlelike bract that looks like a rounded wing above them. After pollination, each flower develops into a rounded capsule that hangs on in groups from the bract and lasts nearly the whole winter. No other trees have anything quite like it.

Four species are commonly found in the region, but none are native (see table 16). At first glance, lindens look very similar to each other, in the same way that pines look similar. But as you get to know them, you'll be able to pick their differences out with ease. None are invasive in our region, so you'll only find them where they have been planted in the landscape.

TABLE 16. QUICK GUIDE TO LINDENS

LEAF	FRUIT	LEAF UNDERSIDE
LITTLELEAF *Tilia cordata* | Common		
2–3 in. (5–8 cm) long, smallest of our lindens; light green below, no contrast between blade and veins, with tufts of orange hairs in vein unions, especially base of midvein	Globes (up to 7), no bigger than ¼ in. (6 mm), velvety; no ribs; short stems	
AMERICAN *Tilia americana* | Common		
4–6 in. (10–15 cm) long, largest of our lindens; light green to whitish below, with small tufts of light-colored hairs predominantly in lateral vein unions	¼ in. (6 mm), globe, round to oval, no ribs; brown, lightly fuzzy; very long peduncle and bract	
SILVER *Tilia tomentosa* | Common		
4–5 in. (10–13 cm) long, dark green above, white and densely pubescent below, creating striking contrast; twigs densely pubescent	¼ in. (6 mm), oval; up to 10 on short, elbowed stems	

LEAF	FRUIT	LEAF UNDERSIDE
EUROPEAN LARGELEAF *Tilia platyphyllos* | Somewhat common		
3–4 in. (8–10 cm) long, green above and below, with heavily contrasting veins below; distinctly fuzzy on leaves, petiole; any tufts of hair yellow, not orange	Prominently ribbed globes (up to 5), greater than ¼ in. (6 mm), lightly velvety; long stems	
CRIMEAN *Tilia × euchlora* | Somewhat common		
2–4 in. (5–10 cm) long, green above and below, with contrasting veins; hairless save for pale tufts in unions of veins below, notably near leaf base	Less than ¼ in. (6 mm), oval, pointed; fuzzy with slight ribs	
MONGOLIAN *Tilia mongolica* | Uncommon		
1.5–3 in. (4–8 cm) long, dark green above, paler below, with large, triangular serrations, nearly lobed	Greater than ¼ in. (6 mm), rounded, pubescent	

Littleleaf Linden
Tilia cordata

This is surely the most common linden species found in the Pacific Northwest. Its name says it all in terms of the best identification characteristic—but you could just as easily encounter a linden tree, blindly guess it's littleleaf, and be right 75 percent of the time. Sometimes it is just that easy.

Large tree up to 70 feet (21 meters) tall, with upright, oval-shaped crown; branches grow up and arch down, giving the branch architecture a fountain-like appearance (often with several upright leaders). **BARK:** Gray, fairly smooth when young, eventually splitting into flat-topped ridges that meander up stem. **TWIGS:** From red to greenish-brown, glabrous with a few lenticels; buds brown to red, *bulbous, often with bulge on one side*, making them look like mittens. **LEAVES:** Smallest of our common lindens, usually around 3 inches (8 centimeters) long, rounded or heart shaped, with abruptly acuminate tip and serrated margins; base often asymmetrical; *tufts of orange hair around vein unions only*, especially around base of midvein. **FLOWERS:** Perfect, yellow, borne in early summer in clusters of 5–7 hanging from paddlelike bract; very fragrant. **FRUIT:** Cluster of small, slightly pointed globes (up to 7); velvety at first *with no ribs*; short stems.

SPECIES REMARKS: Look for the tufts of orange hairs in the vein unions on the underside of small leaves to set this species apart, along with the small fruit clusters and hairless twigs with mitten-like buds. American basswood also has smooth twigs, but they tend to be larger and sometimes covered in bloom.

American Basswood

Tilia americana
Syn. *Tilia heterophylla*

Though not as common as littleleaf, American basswood (also called American linden) can be found as a street tree often enough. Previously it was considered two species, along with the white basswood (*T. heterophylla*), but they've since been merged.

Large tree up to 80 feet (24 meters) tall, with upright, rounded canopy; usually splits into a few large scaffold limbs with smaller limbs arching from them. **BARK:** Similar to other lindens, gray with flat-topped ridges. **TWIGS:** Fairly stout, smooth, light brown to reddish-green, can have whitish bloom; buds plump, broadly conical, with only 2 scales, which gives them lopsided, asymmetrical appearance. **LEAVES:** Largest of our lindens, often up to 6 inches (15 centimeters) long, broadly ovate, with cordate but asymmetrical base and serrated margins; green and smooth above, whiter and smooth below *with tufts of hair primarily along lateral vein unions only.* **FLOWERS:** Perfect, yellow, fragrant, borne in clusters with long (up to 4 inches, or 10 centimeters) paddle-like bract. **FRUIT:** Clusters of ¼-inch (6-millimeter), spherical to oval, slightly fuzzy capsules with no ribs; peduncles longest of lindens.

SPECIES REMARKS: The larger, smooth leaves and longer fruit clusters help to differentiate American basswood quickly. The lighter color of the underside of the leaves makes a striking contrast that you won't see on other species aside from silver linden, but silver linden is densely hairy rather than smooth.

Silver Linden
Tilia tomentosa

Silver lindens are surely the most stately of those planted in the Pacific Northwest. They are found most often in parks but have made it onto a few municipal planting lists, so expect to find them along streets as well.

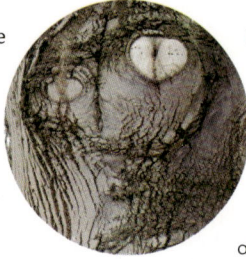

Large tree up to 70 feet (21 meters) tall, with single stem and many upright, outwardly arching limbs, often growing close together; canopy becomes rounded and very dense with leaves. **BARK:** Similar to other lindens, gray with flat-topped ridges. **TWIGS:** Gray, tomentose (covered in downy, white hairs); buds also tomentose, rounded, lopsided due to second bud scale. **LEAVES:** Up to 4 inches (10 centimeters) long, rounded, with cordate, asymmetrical base; large, sometimes double serrations; *highly contrasting between smooth, green topside and white, densely fuzzy underside*; distinctly layered down twigs with flowers and fruit beneath. **FLOWERS:** Perfect, yellow, fragrant, borne in clusters of 7–10, hanging from a paddlelike bract; last of our region's lindens to bloom. **FRUIT:** Collection of ¼-inch (6-mil-limeter) *oval-shaped* capsules on *short, elbowed stems*.

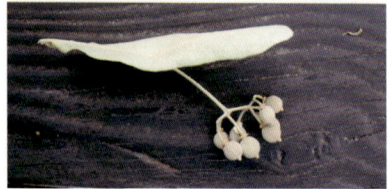

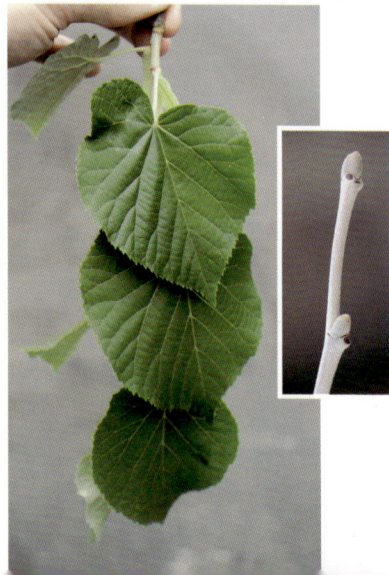

SPECIES REMARKS: The densely pubescent twigs, buds, and leaf undersides are telltale clues for this species. European largeleaf linden also has fuzzy leaves, but they are green on both sides and not as densely covered; it also blooms first rather than last.

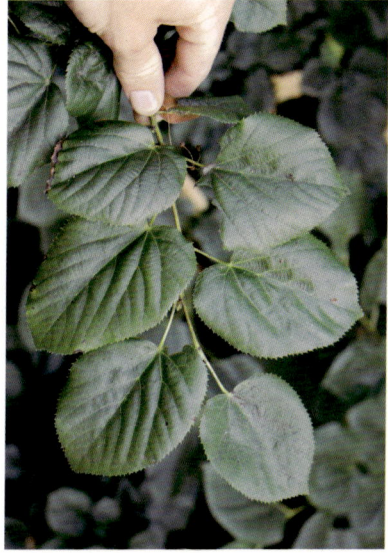

Leaves of Mongolian linden (left), Crimean linden (right), and European largeleaf linden (below)

Similar Species

Mongolian linden (*Tilia mongolica*) is an uncommon species that has small leaves similar to littleleaf's, but they have deep serrations, reminiscent of a stylized drawing of a flame. The serrations can be so pronounced that they look almost lobed. The rounded form stays smaller overall than other species.

European largeleaf linden (*Tilia platyphyllos*) appears to be less common than the others covered here, but then again, it's also easily confused with the littleleaf linden and American basswood, so better to err on the side of diversity. As

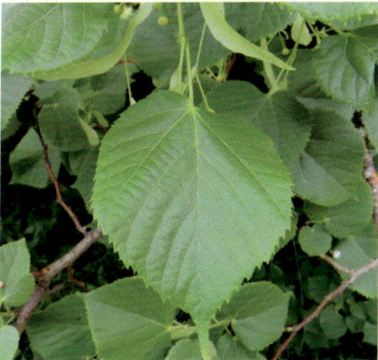

another European species, it's only called "largeleaf" to set it apart from the littleleaf linden (it's "large leaves" are actually smaller than American basswood). Its dark green leaf blade contrasts sharply with its lighter veins, which are distinctly fuzzy along with the rest of the leaf (top, bottom, and petiole). It blooms the earliest of the region's lindens, and its fruit is no larger than ¼ inch (6 millimeters) in diameter, each capsule lightly velvety with five prominent, vertical ribs. Look for the leaf differences to tell it apart, as well as a more buttressed base on mature trees.

Crimean linden (*Tilia × euchlora*) is found sparingly but is certainly confusable when it is. One of its parents is likely littleleaf linden, with which it shares leaf traits, including being mostly hairless save for tufts in vein unions. In contrast to littleleaf, the hairs are pale and leaves larger (up to 4 inches, or 10 cm), with veins that brightly contrast with the leaf blade below. It also has sharp, abruptly serrated margins, and its fruits are oval shaped, slightly ribbed, and covered in a fine layer of dusty wool.

Dove Tree
Davidia involucrata

This ornamental has been planted on the west side of Cascades for some time, but it still isn't found too often. Also called hand-kerchief tree, it's recommended as a street tree more commonly today, and that's where you'll find it most often. **Family: Nyssaceae**

Medium tree up to 40 feet (12 meters) tall, with pyramidal shape when young; with age, loses its dominant central leader, developing a rounded crown. **BARK:** Gray-ish-brown to orange with lenticels; becomes strikingly flaky with age, exfoliating to reveal patches of orange. **TWIGS:** Fairly stout, smooth, mostly brownish, resembling black tupelo, a close relative; buds blackish to red, conical, angle away from stem notice-ably, sometimes grow on short shoots. **LEAVES:** Up to 5.5 inches (14 centimeters) long, broadly ovate with long, acuminate tip; sharply pointed, irregular serrations along margin; veins prominently impressed into leaf blade; base cordate, often overlapping on either side of petiole. **FLOWERS:** Tiny, all crammed together into spherical flower head hanging off long (3-inch, or 8-centimeter) peduncle, sub-tended by *2 large, white bracts*; bracts of 2 different lengths, appearing like ghostly white leaves. **FRUIT:** Large, brown, spherical drupe, over 1 inch (2.5 centimeters) in diameter; persists through fall and early winter.

SPECIES REMARKS: To me, the leaves of the dove tree look intense due to their sharply serrated edges and prominent veins, so they stand out first. When in bloom, the flowers set this tree apart, and in winter, look for the old fruit and unique buds.

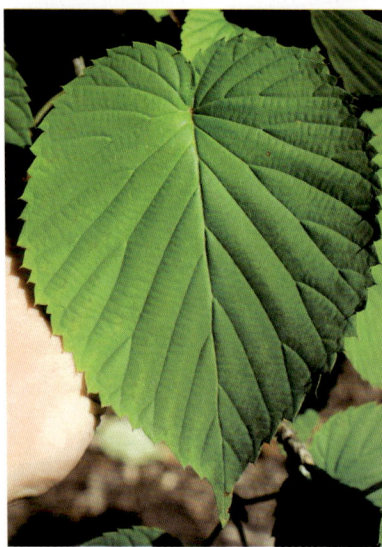

Eastern Redbud
Cercis canadensis

The only simple leafed member of the pea family (see p. 364) in our region, redbuds are classic landscape trees due to their stunning flower displays in early spring. They're routinely found as street or yard trees but don't find their way into natural areas.

Family: Fabaceae

Small tree up to 30 feet (9 meters) tall, with very spherical canopy. **BARK:** Orangish-gray, starting out merely rough, but eventually becoming somewhat shaggy, but not exactly exfoliating. **TWIGS:** Slender, without hairs, usually dark brown to blackish with an obvious zigzag; buds small, black, globe-like or slightly elongated and very scaly; often 2 or more on a node (one vegetative and the others floral); floral buds found all over older limbs too. **LEAVES:** Up to 5 inches (13 centimeters) long, distinctly heart shaped (cordate), often held more or less vertically; veins radiate palmately from base of leaf; *margins entire; petiole has conspicuous swelling just before blade.* **FLOWERS:** Usually pink to purple (one variety is white), pealike flowers that emerge in early spring before leaves, covering the whole tree, appearing often from bare stems. **FRUIT:** Small pea pod, often remaining on tree through winter; borne in clumps or singly along twigs.

SPECIES REMARKS: The pea pods along bare stems and the leaf traits are the tricks for this species. No other pea pod–producing trees have this style of fruiting, called **cauliflory**. Its leaves can be confused with those of katsura, but katsura's are smaller and mostly opposite. **Judas tree** (*C. siliquastrum*) is a related but less common species; its leaves are much more rounded and lack a distinct point but have a cordate base (see below).

TOP LEFT: *Eastern redbud leaf*
BOTTOM LEFT: *Judas tree leaf*

European Smoketree
Cotinus coggygria

This species (also called common smoketree) has been planted more often recently as an option for small planter strips and small front yards. It's more of a shrub most the time, and I have little love for it and its bushy-topped, leggy, unkept habit.
Family: Anacardiaceae

Small tree up to 15 feet (5 meters) tall, with puffy, rounded canopy; often has long, leggy shoots jutting out like solar flares. **BARK:** Brownish-gray with hints of orange; rough, becoming increasingly flaky with age. **TWIGS:** Stout, straight, reddish to purple with contrasting yellow lenticels; buds small, black, pointed, almost clawlike above *unlobed, semicircular leaf scars.* **LEAVES:** Oval to almost round, 1.5–3.5 inches (4–9 centimeters) long, with very long petiole, entire margins, parallel veins; apex usually rounded but can be slightly notched; both green and purple varieties. **FLOWERS:** Borne on big poofy panicles covered in hairs in late spring through early summer; hairs change from yellow to pink with age and are the real show. **FRUIT:** Small, inconspicuous, dry drupe, not wildly helpful.

SPECIES REMARKS: Most parts of this tree help it stand out. The less common **American smoketree** (*C. obovatus*) can grow to be tree sized and shaped: look for the longer, more elliptic leaves (2–5 inches, or 5–13 centimeters, long), more-orange twigs with *lobed leaf scars,* and a single-stemmed, tree-like appearance.

European smoketree bark (left) and American smoketree bark (right)

TOP: *European twig* **BOTTOM:** *American twig* **BELOW:** *American leaf*

Cascara
Frangula purshiana
Syn. Rhamnus purshiana

Native to the moist forests of the region, this little tree has been planted more recently as a small, native tree for streets and small gardens. Also called cascara buckthorn, it's the only tree-sized species of this genus commonly grown here and can be found in natural areas as well as towns. **Family: Rhamnaceae**

Small tree, rarely over 30 feet (9 meters) tall in the landscape, with round canopy usually lacking main leader. **BARK:** Smooth, light gray, with minute striations running vertically; often covered in white lichen. **TWIGS:** Reddish-brown, straight, with soft, russet pubescence; *buds naked*, just a few tiny, immature, folded leaves covered in brown pubescence. **LEAVES:** Up to 6 inches (15 centimeters) long, oval, dark green above, lighter below, with very prominent veins; margin mostly smooth but sometimes with very fine serrations. **FLOWERS:** Small, greenish-white, borne in clusters in leaf axils; appear in early spring, but not wildly conspicuous. **FRUIT:** Pea-sized, jet-black drupe that ripens in fall; apparently very sweet.

SPECIES REMARKS: Cascara would be easy to confuse with serviceberries (*Amelanchier* spp.) if it wasn't for the buds, which should be the first and last clue you need. The leaves also stand out once you see them due to their prominent veins.

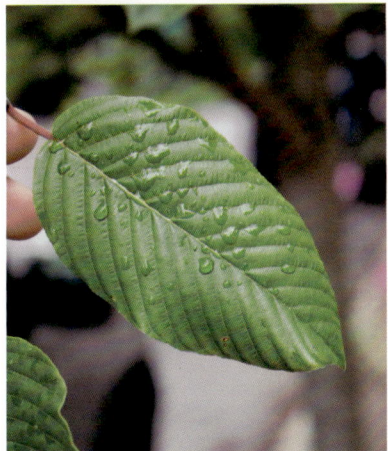

Chinese Persimmon
Diospyros kaki

Two persimmons can be found in this region, mostly west of the Cascades, but Chinese persimmon (also called Japanese persimmon) is most common. It's found sparingly as a street tree, more often in the landscape as a small fruit tree.

Family: Ebenaceae

Small tree up to 30 feet (9 meters) tall, with shaggy, rounded canopy; tends to remain somewhat bushy, but with good height. **BARK:** Light brown to gray, flaky, exfoliating to reveal new bark below. **TWIGS:** Yellowish-brown with slight, rough pubescence; buds reddish-brown, conical, somewhat flattened. **LEAVES:** Up to 7 inches (18 centimeters) long, ovate, deep glossy green on top, lighter with pubescence below; margins entire; often hang downward or look slightly unkempt. **FLOWERS:** Dioecious; male trees produce pollen from small, whitish flowers in groups of 3; female flowers larger, solitary, subtended by 4 distinctive bracts. **FRUIT:** Persimmon, green turning orange at maturity; can either be egg shaped or a squished sphere, like a baby pumpkin but smaller and smooth skinned; *always has crown of 4 bracts.*

SPECIES REMARKS: The big, glossy leaves are often the first clue to this species, but the flaky bark and the fruit with 4 bracts certainly set it apart. You may stumble upon **American persimmon** (*D. virginiana*), which differs in three main ways: its bark is dark gray, hard, and broken up into rectangular blocks; it grows much taller (up to 50 feet, or 15 meters); and its fruit is much smaller, only 1.5 inches (4 centimeters) in diameter.

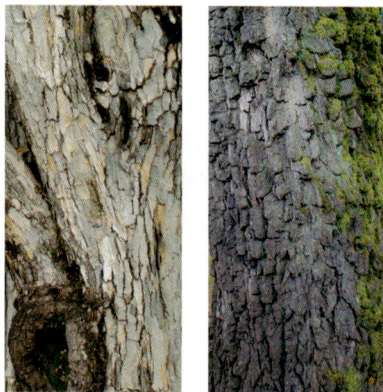

Bark of the Chinese persimmon (left) and American persimmon (right)

Black Tupelo
Nyssa sylvatica

This tree has been planted like gangbusters over the last few decades, especially in cities west of the Cascades. It's almost always a street tree and never found in natural areas. Of all the overplanted trees today, I like this one the best. **Family: Nyssaceae**

Medium tree up to 50 feet (15 meters) tall, with upright yet rounded, layered canopy; often has a single stem or a few main leaders, with dense complement of horizontal twigs; canopy often feels busy. **BARK:** Starts smooth, light gray but becomes at first rougher, darker, then ultimately intensely fissured; very similar to American persimmon or sourwood (see sidebar, p. 262). **TWIGS:** Light brown, smooth, often with several short shoots; chambered pith; buds ovoid, pointed; *leaf scars below buds have 3 distinct vascular bundle traces* markedly indented into scar. **LEAVES:** 3–6 inches (8–15 centimeters) long, ovate to obovate (widest section nearer tip); glossy, waxy green on top and duller beneath; *margins entire.* **FLOWERS:** Dioecious; both produced in inconspicuous clusters, but males look like pom-poms. **FRUIT:** Oblong blue drupes on long peduncles; appear in clusters but can seem solitary if only 1 flower is fertilized.

SPECIES REMARKS: Once you recognize the horizontal, layered silhouette of black tupelo, they will stick out to you at any time of year. The fall color is a striking red, contrasting nicely with the blue drupes.

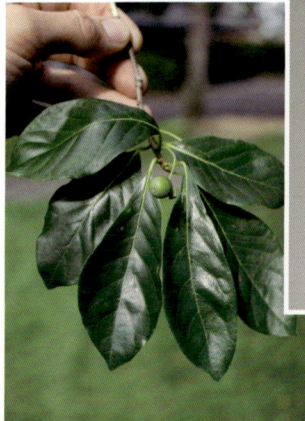

TWO TUPELO LOOK-ALIKES

If you see a tree that looks like a black tupelo, odds are it *is* a black tupelo based on how common they are alone. But if something is just a little off (and you've already ruled out persimmons and dove tree, right?), then you may have a sourwood or a sassafras on your hands. Both have blocky bark and similarly shaped leaves to black tupelo's, but luckily they have some striking differences too.

Sourwood (*Oxydendrum arboreum*) is a small tree with a spindly, droopy appearance. Its leaves differ from tupelo's by having a more acuminate tip and small serrations along the margin (black tupelo has entire margins). As its name indicates, if you pick a leaf and give it a chew, it'll taste tangy, sour, and acidic, but not entirely bad. Sourwood is in the heath family, Ericaceae, and produces clusters of small, urn-shaped flowers that mature to dry capsules, far different from the tupelo (and from other look-alikes for that matter).

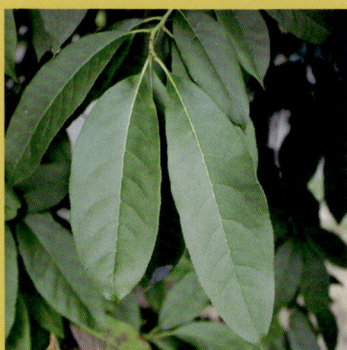

Sassafras (*Sassafras albidum*) is a medium tree that is less spindly than the sourwood but not nearly as dense as the tupelo. Its leaves are slightly wider than those of most look-alikes, but more importantly they can grow lobes. Sassafras leaves can have no lobes, a single lobe reminiscent of a mitten, or two symmetric lobes; often you can find all three on a single twig, but certainly with some perusing you can find this variation throughout the tree. Like the tupelo, they have bluish-black drupes; however, each is borne more or less singly with a bright red receptacle.

CLOCKWISE FROM LEFT: *Saucer magnolia in bloom; deciduous magnolia leaf; 'Butterflies' magnolia flower*

MAGNOLIAS

Magnolia

A storied genus that is nearly universally adored, magnolias (family Magnoliaceae) are about as common as you can get west of the Cascades. Though they're less common east of the Cascades, you'll still find them planted in relatively protected or otherwise less extreme areas. How well they are doing there may be a different story because they tend to prefer more moisture in summer.

There are a lot of species of magnolia (more than 350!), mostly from Southeast Asia, but several come from southeast North America as well. As is the case for florally endowed species, they have been bred, crossed, and hybridized to oblivion and back again—it's almost impossible to assign the commonly planted deciduous trees to a given species. As is the case elsewhere in this guide, I am avoiding it as best I can. *The Tree Book* by Dirr and Warren can help shed light for those who want to go deeper (see the resources in the back of the book).

I focus on the genus-level characteristics for the deciduous species so that you can at least confidently say, "That's a magnolia." Then, I've listed traits to look for to separate out some of the more common species. The evergreen species you'll nail down, no problem.

Deciduous Magnolias

Deciduous magnolias range from small, multistemmed shrubs to large trees, and nearly all can be interbred to get a wild amount of variation in their forms and flowers (with a heavy focus on the flowers). However, they tend to have a few traits in common that at least let you know you are looking at a magnolia. You'll find them all in the landscape but never in natural areas.

Form: With the exceptions of cucumber magnolia and the few giant-leafed species that grow quite tall with a single stem, most of the treelike magnolias develop rounded, spreading crowns, often with multiple scaffold limbs and no real central leader.

Bark: Mostly smooth and light gray. Only with old age will they develop splits and some furrows. Cucumber and umbrella magnolia are the exceptions in that they develop coarser bark with shallow ridges and furrows.

Twigs: Most twigs are stout, brittle, and widely spaced, tipped with a comically large, densely fuzzy terminal bud, a lot like a rabbit's foot (remember those?). Twigs appear jointed or armored due to their large leaf scars and marks that encircle the twig.

Leaves: Mostly oblong or obovate, so they are usually widest nearer the middle or toward the tip. They have entire margins, pointed tips, and usually some amount of pubescence on them. To be honest (and no offense meant), I think they have the most generic of leaves.

Flowers: Large, showy, and made up of tepals—essentially undifferentiated sepals and petals. Their stamens and pistils are all bunched in the center on a cone-shaped structure that becomes the fruit.

Fruit: A unique, lumpy, conelike aggregate of follicles, each with a single, often bright red seed inside that tends to be exserted at maturity for birds to eat. Sometimes the whole fruit stays brown and woody, sometimes it falls away before maturing, and sometimes it's a bright red, lumpy mass, but there's really nothing else like it.

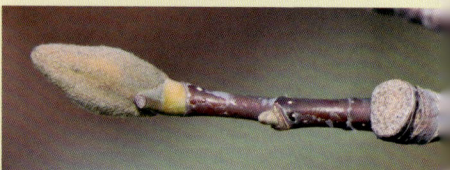

Star magnolia flower and leaf

Star magnolia (*M. stellata*) is a small, shrubby species, often with multiple stems and a somewhat contorted form. The leaves are the smallest of the region's common species, up to 4 inches (10 centimeters) long, tapering from a wide tip down to a thin base, usually with a rounded apex. It's the first of the magnolias to bloom in early spring; the flowers are made up of *over a dozen* thin, white, strap-like tepals that arch outward in a messy spray.

Kobus magnolia (*M. kobus*) is a small tree, rarely over 30 feet (9 meters) tall, with an upright, pyramidal canopy. Its leaves are up to 6 inches (15 centimeters) long and broadly ovate with a sharply acuminate tip. The flowers are white like star magnolia's but have only *six to nine* tepals, which are also wider (see below).

Saucer magnolia (*M.* × *soulangeana*) is the most common deciduous magnolia in our region. It's a hybrid between yulan magnolia (*M. denudata*) and lily magnolia (*M. liliiflora*) and is itself a likely parent of hundreds of other varieties and hybrids. It's a low-spreading tree, up to 20 feet (6 meters) tall but easily 30 feet (9 meters) wide or more. Its leaves are 6 inches (15 centimeters) long, obovate with an abruptly acuminate tip. The flowers look like tulips when they open, with usually nine wide, rounded tepals that range from white with purple tinges to entirely pinkish.

LEFT: *Kobus magnolia flower*
RIGHT: *Saucer magnolia leaf and flower*

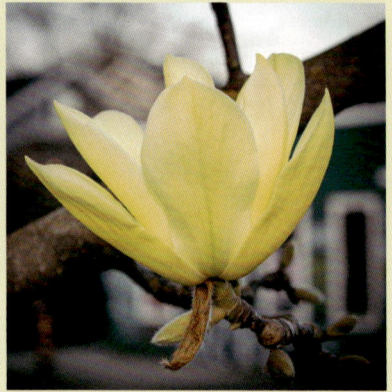

Cucumber magnolia (*M. acuminata*) is the biggest and tallest species you'll find around here, able to reach nearly 80 feet (24 meters) tall. Though it's a common parent of many hybrids, it's somewhat rare to find the true species. Its flowers are small, strappy, and mostly green, barely able to be seen with a keen eye. The leaves can get up to 10 inches (25 centimeters) long and are standardly oval with a pointed tip; the bark is ridged, which is a distinctive feature.

Hybrid magnolias (*M.* [enter catchy name]) make up the hordes of ornamental cultivars bred from the species above and many others, each differing in flower color, tepal width, tepal number, growth habit, and the like. Some common ones planted out here are 'Butterflies,' 'Elizabeth,' and 'Galaxy' the first two being quintessential yellow varieties and the last a purple one. Any further discussion I will leave to the nursery-trade junkies.

CLOCKWISE FROM TOP LEFT: *Cucumber magnolia bark and leaf; 'Butterflies' flower; 'Elizabeth' flower; 'Galaxy' flower; 'Elizabeth' leaf*

TOP: *Umbrella magnolia flower and leaf* **BOTTOM:** *Bigleaf magnolia leaf and flower*

Umbrella magnolia (*M. tripetala*) and **bigleaf magnolia** (*M. macrophylla*) are somewhat unique finds on the west side of the Cascades, but you'll know them when you see them. Each getting to over 30 feet (9 meters) tall, they both have massive leaves, with umbrella's getting to 2 feet (60 centimeters) long and bigleaf's to 2.5 feet (80 centimeters). They each also have giant, creamy white flowers, nearly 10 inches (25 centimeters) across, that are fragrant (to some people, unpleasantly so for umbrella magnolia). The leaves, flowers, and fruit set them apart: Umbrella magnolia has a gently tapering, wedge-shaped leaf base; an earlier-blooming flower; and a thin, reddish fruit. Bigleaf magnolia has a heart-shaped leaf base with little earlobes on either side of the petiole, purple spots often at the base of the inner tepals, and plumper, brown fruit (see photo on p. 264).

Southern Magnolia
Magnolia grandiflora

This is the most common evergreen magnolia in our region—and really anywhere else too. It's most commonly planted as a street tree nowadays, but historically it was a fashionable landscape tree, so you can find it in both spaces.

southern magnolia apart from all others. **Sweetbay magnolia** (*M. virginiana*), another evergreen species commonly found, has smaller and thinner leaves that are white, glaucous, and minutely pubescent below.

Medium evergreen up to 60 feet (18 meters) tall, with upright but rounded habit; tends to be vertically oriented, coarsely textured. **BARK:** Brownish, smooth but bumpy when young, becoming irregularly flaky, rough with age. **LEAVES:** Large, leathery, up to 10 inches (25 centimeters) long; dark green above, *distinct orange fuzz below*; fuzzy twigs are tipped with a large flower bud or smaller, brown vegetative bud. **FLOWERS:** Perfect, large, usually around 8 inches (20 centimeters) across but can be wider; usually 6–12 broad, white tepals; flowering peaks in midsummer, tapering off through fall. **FRUIT:** Brown aggregate of follicles; each with a red seed.

SPECIES REMARKS: The broad evergreen leaves with the orange fuzz beneath will set

BELOW: *Sweetbay magnolia flower and leaves*

STEWARTIAS

Stewartia

It is my opinion that this genus (family Theaceae) is underrepresented in our land-scapes. Though it's far from unknown, it is simply overshadowed by the dogwoods and cherries, whose blooms are more profuse compared to the stewartias' more subtle dis-plays. They are worth seeking out though.

Three species are commonly found planted in the Pacific Northwest, mostly west of the Cascades, as they prefer rich, moist soils, tending to not tolerate those that are too dry, mineral-rich, or compacted (see table 17). A fourth species, Korean stewartia (*S. koreana*), is often listed; however, it's been found to be essentially no different from Japanese stewartia (*S. pseudocamellia*; see below) and thus has been reclassified with it.

These are landscape trees mostly, but they can be found planted as street trees (where more often than not they are struggling mightily). They don't seed themselves in, so you won't find them in wild areas.

TABLE 17. QUICK GUIDE TO STEWARTIAS

SPECIES & KEY TRAITS	BARK	BUDS
Japanese *S. pseudocamellia* Leaves 2–3.5 in. (5–9 cm) long, widest of common species; bark camouflage patterned; flower large; fruit lacks bracts		
Tall *S. monadelpha* Leaves 1.5–3 in. (4–8 cm) long; leaves, flowers, fruit, smallest of common species; bark orange, flaky		
Rostrata *S. rostrata* Leaves 3–6 in. (8–15 cm) long, sharply serrated; bark gray, rough; flower large; fruit has 5 bracts		

Japanese Stewartia
Stewartia pseudocamellia
Syn. *Stewartia koreana*

Also called Korean stewartia, this is the most common of the three stewartia species in the Pacific Northwest, with something for everyone no matter the time of year. If only they made a delicious edible fruit, I bet they would be voted best overall small tree.

Small tree up to 30 feet (9 meters) tall, with upright, slightly rounded canopy; usually several main limbs that all grow upward and outward. **BARK:** Smooth; exfoliating in large, irregular puzzle-piece-like patches, leaving a mosaic of browns, beiges, dark reds; visually striking even on young trees. **TWIGS:** Reddish-brown, mostly lacking lenticels; buds moderately long, pubescent, flattened, with pointed tip; few imbricate scales. **LEAVES:** Usually up to 3.5 inches (9 centimeters) long, somewhat glossy, with veins set into blade; tip acuminate mostly, but can be rounded; margins sparsely serrate or crenate. **FLOWERS:** Perfect, white with orange anthers, 2 inches (5 centimeters) in diameter, cup shaped; petals wrinkly, delicate looking, emerge from spherical bud. **FRUIT:** Starts out as fuzzy sphere, like the flower bud, then dries into sharply pointed, brown capsule that splits into 5 valves; stays on tree through winter; has either no bracts or strongly reflexed bracts. **SPECIES REMARKS:** The bark and fruit will be visible all year and give Japanese stewartia away; no other tree has similar fruit, and the bark is similar only to kousa dogwood's, which has oppositely arranged buds. The flowers should also set this species apart in early summer.

Similar Species

Tall stewartia (*S. monadelpha*) has gorgeous bark that is a bit rougher than Japanese stewartia's, exfoliating in smaller patches with a more orange hue overall. It also has the smallest leaves, flowers, and fruits of our common species, so it comes off as a miniaturized version of the others. Its buds look like little flattened cones or grass flowers, covered in very fine pubescence.

Rostrata stewartia (*S. rostrata*) differs from the others by having rougher, gray bark that develops small fissures and furrows (rather than exfoliating in patches); even youngsters of the other two species develop exfoliating bark. Its leaves are glossier, and its fruit has a distinct beak throughout its development. It also generally retains the bracts below the fruit, giving it a flowerlike appearance. The scientific name rostrata (sounds like "rose") means "beaked".

ABOVE: *Tall stewartia leaf, flower, and fruit* **BELOW:** *Rostrata stewartia flower, leaf, and fruit*

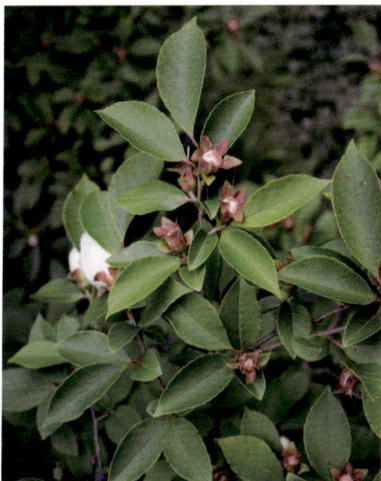

Japanese Snowbell
Styrax japonicus

Another common ornamental flowering tree that won't bother you by getting too big, Japanese snowbell is found as a landscape and street tree on the west side of the Cascades. Its showy displays in spring are stunning. It's not invasive, so you'll only find it where planted. **Family: Styracaceae**

Small tree up to 25 feet (8 meters) tall, with rounded, compact habit. **BARK:** Light brownish-gray, generally smooth but some shallow striations and furrows; often develops bulbous growths with age. **TWIGS:** Slender, light brown, slightly zigzag; buds technically naked, brown, finely tomentose; *almost always in pairs with smaller bud piggybacked on large one, appressed to twig.* **LEAVES:** Oblong, usually no longer than 3 inches (8 centimeters); tip tapers to point; margin very slightly dentate if at all, generally appearing entire. **FLOWERS:** Perfect, small, white with yellow stamens, profuse in midspring, with slight fragrance; borne on tiny shoots in leaf axils along twigs and hanging below leaves, giving tree layered appearance. **FRUIT:** Dry, nearly spherical drupe attached to pedicel by a little scalloped cap (very cute); as profuse as flowers and stays through winter.

SPECIES REMARKS: The piggybacked buds will give Japanese snowbell away every time in winter. The bark also seems to stand out once you recognize it (though it's fairly drab overall); the flowers and fruit are always good clues.

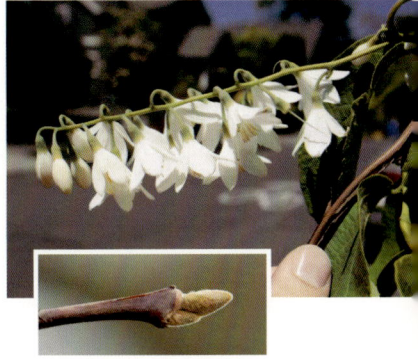

Similar Species

Bigleaf snowbell (*Styrax obassia*), as its name suggests, has big leaves, up to 8 inches (20 centimeters) long, far larger than Japanese snowbell's. Its leaves also have more prominent points along the margin, especially toward the tip, and the base of the petiole covers the buds. The flowers and the fruit are very similar to Japanese snowbell's except they are borne on long racemes that can hang around through late fall. The long chains of fruit, along with the overall larger form (up to 30 feet, or 9 meters) and wider branch spacing, separate this species nicely. Like its cousin, it has fuzzy buds that are also piggybacked, but they are much larger, more yellow, and grow away from the purply twigs.

Carolina silverbell (*Halesia carolina*) can grow larger than either of its kin (up to 40 feet, or 12 meters), but when young its canopy is more like an oversized Japanese snowbell. Its leaves are a bit larger than Japanese snowbell's (up to 4 inches, or 10 centimeters, long) and finely serrated, looking more like a stewartia than a snowbell. The flowers are borne in small clusters of two to five and hang down from the twigs in late spring. In fall, small, papery, four-winged fruits develop and can hang on through the winter. The buds are single, conical, and have imbricate scales. Look for the buds and remnants of the fruit in winter.

TOP ROW: *Bigleaf snowbell fruit, leaf, flower, and twig* **RIGHT COLUMN:** *Carolina silverbell flower, leaf, fruit, and twig*

THE ROSE FAMILY

The rose family (Rosaceae) is a big one; it includes plenty of species you wouldn't expect at first glance because they differ dramatically in their fruit, from the strange, tailed achenes of mountain-mahogany (*Cercocarpus*), to the seeds we call almonds (*Prunus amygdalus*). What pulls them all together is their flowers, which have five sepals and five petals that all fuse at the base to create a **hypanthium**, or a cupped base. You'll notice that in all the genera below, no matter how their flowers are arranged (i.e., groups or singly), they all look like slight variations on this one design.

The best way to differentiate these genera is to familiarize yourself with their small details. Below is a quick summary of the broad traits that you can use to at least narrow down your tree to a genus. But remember, these are broad traits, and you'll always encounter a bit of variation, so take this as a guide, not gospel.

TABLE 18. QUICK GUIDE TO THE ROSE FAMILY

SPECIES & KEY TRAITS	FRUIT	BARK
Serviceberries *Amelanchier* Leaves small, oval, with minute serrations; twigs smooth, skinny, buds pointed, lack thorns; flowers usually white, strappy, borne in upright racemes, petals thin, long; smooth bark, vertical lines		
Hawthorns *Crataegus* Leaves hairless, serrated margins, often lobed; twigs distinctly thorny; flowers white, borne in profuse, flat-topped clusters; flaky, scaly bark		
Rowan, mountain-ash, whitebeam *Sorbus, Aria, Alniaria* Leaves pinnately lobed (*Aria, Alniaria*) or compound (*Sorbus*); twigs tomentose, stout; flowers white in dense, flat-topped clusters; no thorns; smooth bark		

Pears

Pyrus

Leaves not pubescent, margins mostly entire, petiole long; twigs with fewer hairs than apples'; flowers white, in compact clusters, smell bad; rough, ridged bark

Apples

Malus

Common apple: leaves pubescent below with serrations, twigs hairy; Crabapple: leaves glabrous, twigs hairy or not; flowers white, pink, or red, profuse, smell quite sweet; smooth, scaly bark

Cherries

Prunus

Leaves hairless, margins serrated, glands on petiole/leaf base; twigs hairless with strong dimorphic tendencies; flowers white to pink, borne in clusters (not singly), often with notches in petals; smooth bark with horizontal lines

Plums

Prunus

Leaves papery, slightly serrated; twigs slender, dark black, lack spurs, sometimes end as sharp thorns; flowers white or soft pink, borne singly, bloom earliest; rough bark

Crabapples have quintessential rose family flowers, and they're some of the loveliest to boot.

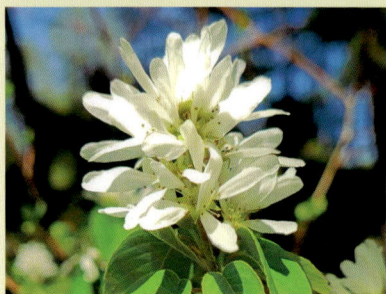

CLOCKWISE FROM TOP LEFT: *Serviceberry fruit, bark, and flower*

SERVICEBERRIES

Amelanchier

As a genus, serviceberries can be picked out fairly easily by their sprays of white, strappy flowers in early spring, their cute little leaves, and their blueberry-like fruit (which tends to get infected by a disease called rust, appearing as if they are covered in orange mold). But sorting out which species is which can be challenging, even for the best dendrologists.

Three species and one hybrid are found in the landscapes of the Pacific Northwest. The native western serviceberry, or saskatoon, isn't commonly found as a landscape tree but can be found as a thicket-like tree in natural areas. If you see a well-formed street tree, it's likely an Allegheny or downy serviceberry, both native to eastern North America, or a selection of their natural hybrid, an apple serviceberry.

Incidentally, it's a safe bet that you won't see much shadblow or Canadian service-berry (*Amelanchier canadensis*) in the region. Reportedly, what is often listed in the nursery as *A. canadensis* is in fact one of the species covered here or the hybrid; shad-blow is a multistemmed, shrubby species, definitively different in form from the closely related, eastern, more treelike species.

I can say with confidence that the Allegheny and downy serviceberries are well represented in our cities and towns. However, to most easily tell them apart, you need to find them in spring and examine their emerging leaves and the flowers that follow.

276

Western Serviceberry
Amelanchier alnifolia

The Pacific Northwest's native species, this serviceberry grows wild throughout the region mostly as an edge shrub or a wild-looking, multistemmed, treelike "thing." It is occasionally planted as a street and landscape tree, especially in native-oriented plantings. If you find a serviceberry growing in a natural area, it is almost certainly this one.

Small tree up to 30 feet (9 meters) tall but all too often shorter in the landscape; if intentionally planted, appears as a single-stemmed tree, often with some suckers; otherwise, multistemmed with a lot of suckers. **BARK:** Smooth, gray, with darker vertical striations running up stem; older trees develop shallow cracks. **TWIGS:** Brown, slightly angular, with sparse lenticels and thin, gray epidermis layer that peels off; buds conical, pointed, with imbricate scales, usually no more than ½ inch (13 millimeters) long; scales have silvery hairs coming from their edges. **LEAVES:** 1–2.5 inches (2.5–6 centimeters) long; oval or round, usually with rounded tip, but could be slightly pointed or notched; slightly pubescent when young, but hairless with age; *coarsely serrated, usually only on upper half of margin.* **FLOWERS:** Perfect, white, with thin, widely spaced petals creating a star shape; borne on tight, erect racemes not much more than 1 inch (2.5 centimeters) tall. **FRUIT:** Berry-like pome, green in spring, changing to red and finally nearly black when mature; often marred by rust disease.

SPECIES REMARKS: In summer, look for the smaller, rounder leaves with coarse serrations concentrated near the tip; others' leaves are longer and more pointed, with finer serrations. In spring, look for very compact flower racemes that dot, rather than blanket, the tree.

LEFT: *Downy serviceberry flowers* RIGHT, TOP: *Downy serviceberry fruit* RIGHT, BOTTOM: *Emerging leaves of Allegheny serviceberry (left) and downy serviceberry (right)*

Similar Species

Downy serviceberry (*A. arborea*) is the most treelike species, often growing to 25 feet (8 meters) tall, with a rounded canopy and olive green to reddish twigs with pointed buds. Its leaves are 1–3 inches (2.5–8 centimeters) long and oval, with an abruptly pointed tip, usually a cordate base, and very fine serrations along the margins. When the leaves, shoots, and flowers first emerge, they are green (not red) and *densely covered with gray hairs*, giving them a silvery appearance. The ovate-petaled flowers are white and emerge about a week earlier than the Allegheny serviceberry's, on nodding 2–4 inch (5–10 centimeter) long racemes.

Allegheny serviceberry (*A. laevis*) looks similar to downy serviceberry in most ways, but its twigs are more reddish-brown and its leaves are hairless, with a longer, more tapering tip, a rounded base (usually not quite cordate), and more coarsely and sparsely serrated margins. When the leaves first emerge, they are *distinctly reddish tinged and almost completely hairless*. The white flowers have thinner, paddle-shaped petals and are borne on slightly longer (up to 5 inches, or 13 centimeters), droopier racemes, giving the whole inflorescence a looser appearance altogether.

Apple serviceberry (*A. × grandiflora*) is tougher to categorize because different lineages or selections may resemble one parent or the other to differing degrees; therefore, you can't always say for sure if it's the species or a hybrid. Suffice it to say that if you find a serviceberry that doesn't quite match the descriptions for the native species or the two eastern species and it has pinkish flowers, it's probably an apple serviceberry.

BELOW: *Allegheny serviceberry leaves and flowers*

CLOCKWISE FROM TOP LEFT: *'Paul's Scarlet' midland hawthorn flowers; English (single-seed) hawthorn flowers; cockspur hawthorn twig; Lavalle hawthorn leaves*

HAWTHORNS

Crataegus

Mostly planted as street or yard trees, hawthorns are some of the most common trees in the landscape of the Pacific Northwest, owing much of their popularity not only to their robust flower displays but also to their ties to old European gardens. The European species have been cultivated for generations and helped usher in an easy appreciation of some especially lovely North American species.

Only one native species to the Northwest, black hawthorn, makes it to tree size and into cities and towns, though it does so less often than ornamental or invasive species that escaped cultivation. It's common in more wild areas, and you'll likely start seeing it more as people opt for native alternatives in their gardens and street planter strips.

The flaky bark; dense, crisscrossing, twiggy canopies; and thorns of the common landscape hawthorns simultaneously unite them as a genus and separate them from other rose family look-alikes. On top of that, you know you have a hawthorn when the tree exemplifies its own name: "haw" is the old term for the fruit, and they are absolutely covered in thorns, the only rose family tree genus in our area with so many.

Conveniently, the name also sets different species apart: to quickly narrow down your species of hawthorn, just look at the haws and the thorns (see table 19). Each species has a slightly different combination of thorn and fruit traits, so you'll eliminate unlikely species quickly by focusing on those first. Then confirm the species by considering the flower timing and leaf characteristics, and you're bound to succeed.

TABLE 19. QUICK GUIDE TO HAWTHORNS

SPECIES & KEY TRAITS	HAWS	THORNS (TWIGS)
English (singleseed) *C. monogyna* Leaves up to 2 in. (5 cm) long, dull green, 3-5 deeply cut lobes; thorns short, less than 1 in (2.5 cm); flowers white, 1 style, first to bloom in spring; small red haws, 1 seed; common, invasive		
Midland *C. laevigata* Leaves up to 2 in. (5 cm) long, usually 3 shallow, toothed lobes, often defoliated by July; thorns short, less than 1 in. (2.5 cm); pink-flowered types found; small red haws in loose clusters; semi-common		
Lavalle *C. × lavalleei* Leaves up to 4 in. (10 cm) long, dark green, leathery, lustrous, unlobed; thorns stout, 2 in (5 cm) long; flowers white, in big flat-topped clusters; haws red, second largest, in dense clusters; common		
Washington *C. phaenopyrum* Leaves 2.5 in. (7.5 cm) long, triangular, two basal lobes, serrated margin; thorns long, thin, 1-3 in. (2.5-8 cm); flowers white, in terminal clusters, last to flower; haws small, red, densely clustered, persist all winter; common		

SPECIES & KEY TRAITS	HAWS	THORNS (TWIGS)
Cockspur *C. crus-galli* Leaves 1-4 in. (2.5-10 cm) long, leathery, unlobed, finely serrated margins; thorns up to 3 in. (8 cm) long, often branched (thornless varieties also); flowers in white clusters; haws muted red; common		
Black *C. douglasii* Leaves up to 2 in. (5 cm) long, entire margin near base, then serrated, then lobed; thorns over 1 in. (2.5 cm) long, substantial; flowers early, white, in dense clusters; haws reddish-purple then black; semi-common		
Winter King Green *C. viridis* 'Winter King' Leaves up to 2.5 in. (6 cm) long, like black hawthorn's but more distinctly lobed and triangular; thorns thin, up to 1 in. (2.5 cm), often absent; flowers white, in flat clusters; haws red, in loose clusters; uncommon		
Autumn Glory *C.* 'Autumn Glory' Leaves up to 3 in. (7 cm) long, glossy green with 3-5 serrated lobes per side; thorns short, less than 1 in. (2.5 cm) often absent; flowers white, 2 styles; haws largest, bright red, glossy; uncommon		

English Hawthorn
Crataegus monogyna

This is the most common hawthorn (fittingly, also called common hawthorn) in the Northwest because it's the most invasive. If not found as a street tree, you'll find it growing in dense thickets along ignored landscape edges and vacant lots or popping up in natural areas.

Small tree up to 30 feet (9 meters) tall, with dense, twiggy, rounded habit; branches tend to interweave and sprout vertical shoots. **BARK:** Light gray, broken up into many small, thin, vertical flakes that exfoliate, leaving nut-brown patches beneath; older stems with seams, small burls, thorns. **TWIGS:** Gray-brown, slender but dense, armed with plenty of short thorns up to 1 inch (2.5 centimeters) long, usually slightly smaller; buds small, red, domed, often pointed roughly perpendicular from twigs; can be slightly pointed on newer twigs. **LEAVES:** Around 2 inches (5 centimeters) long, usually with 5 *deeply incised* lobes (more than halfway to midrib), sometimes 3 or 7 lobes; serrated margins and yellow fall color. **FLOWERS:** Perfect, white, with 5 petals, many stamens, and only *1 pistil*, borne in rounded clusters (corymbs), appearing in early spring. **FRUIT:** Haws are small, round, red pome, about ⅜ inch (1 centimeter) in diameter; *single seed inside yellowy pulp* (midland hawthorn, see next page, has 2–3 seeds).

SPECIES REMARKS: The small fruit and small, deeply lobed leaves set this species apart from other hawthorns, except midland hawthorn. Don't confuse it with escaped cherry plums (*Prunus cerasifera*; see the plums section), which can also have thorny growth and spring blooms; the plums' flowers are borne singly (not in clusters) and appear far earlier in spring.

A HAWTHORN DOPPELGANGER

Midland hawthorn (*Crataegus laevigata*), which is also sometimes confusingly referred to as English hawthorn, is commonly found as a street tree in the Pacific Northwest. It's easily confused with its closely related cousin, the other English (singleseed) hawthorn, and for good reason: they look nearly identical. Their size, habit, bark, and twigs are essentially interchangeable, down to the size of their thorns (though midland hawthorn tends to have fewer thorns overall).

To differentiate them, look at their leaves and flowers: midland hawthorn's leaves are slightly shorter (less than 2 inches, or 5 centimeters) and usually have just *three shallow lobes* with serrated margins; they can have five messy-looking lobes, but they aren't nearly as deeply incised (less than halfway to the midrib for midland).

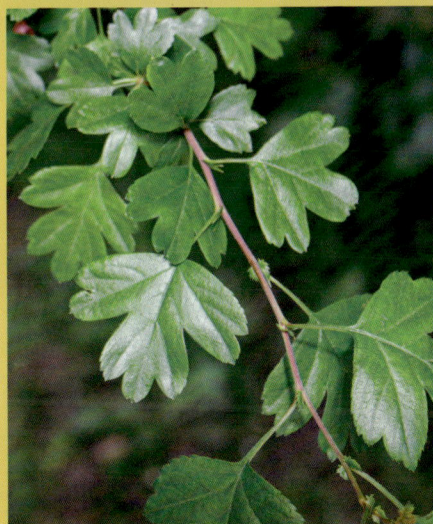

'Crimson Cloud' flowers and leaves

Midland hawthorn was popular due to its profuse pink-flowering cultivars 'Crimson Cloud' and 'Paul's Scarlet', which are the only examples I've seen planted in the Pacific Northwest, and which easily set the species apart in spring. The true species form has white flowers similar to English (singleseed) hawthorn but differs by having two or three styles rather than one. Also look for its lack of leaves and fruit in midsummer and fall: most of the leaves fall away by August due to a fungal infection, and the more common 'Paul's Scarlet' cultivar with double flowers (see genus introduction, p. 279) doesn't produce any haws.

Lavalle Hawthorn

Crataegus × lavalleei

The most handsome of all our hawthorns, Lavalle is the star of the group. Luckily, it's also quite common and potentially the easiest to identify at any time of year. Not invasive, it's found most commonly as a street tree.

Small tree up to 30 feet (9 meters) tall, with very dense crown; it grows a pillowy form, like a cumulus cloud made of leaves. **BARK:** Dark gray, appearing flaky, similar to other hawthorns, but flakes are more irregularly shaped and reflex back rather than fall away, revealing darker brown beneath; *stems almost always spiraled* with many seams. **TWIGS:** Reddish-brown, thicker than others, *impressively stout thorns*, 2 inches (5 centimeters) long; buds red, dome-like with small imbricate scales. **LEAVES:** 2–4 inches (5–10 centimeters) long, dark, lustrous green, *leathery, unlobed*; margins serrate but often only on upper half of leaf; often stick around into late fall. **FLOWERS:** Perfect, white, in flat-topped clusters appearing in late spring, one of the last hawthorns to bloom; striking contrast with dark leaves. **FRUIT:** Haws grow in clusters of several large (¾ inch, or 2 centimeters, long), rounded to oblong pomes; start orangish, mature to dark red in late fall; second-largest haws of Northwest's common species.

SPECIES REMARKS: Many traits set this species apart. The longer, lustrous green, unlobed leaves; the long, stout thorns; the spiraled, somewhat scaly stems; and the big clusters of red fruit on dark green leaves are all unique to this hybrid.

Washington Hawthorn
Crataegus phaenopyrum

Native to eastern North America, Washington hawthorn is widely used in the nursery trade, being sold as a street tree and small landscape tree all over the place. It's found only as a planted landscape tree.

Small tree up to 30 feet (9 meters) tall, with rounded, twiggy yet airy habit. **BARK:** Light gray on the outside, often breaking off in rectangular flakes, revealing handsome orange-brown below; stem usually not twisted. **TWIGS:** Like other hawthorns, brownish with red, domed buds; has many uniformly spaced thorns, *each very long (1–3 inches, or 2.5–8 centimeters), slender, deathly sharp.* **LEAVES:** Up to 2.5 inches (6 centimeters) long, more *distinctly triangular than others, with 2 prominent basal lobes* and 1 central lobe with rough serrations; usually lacks stipule at base of petiole. **FLOWERS:** White, borne in terminal, rounded, pom-pom-like clusters; *last hawthorn to flower.* **FRUIT:** Haws grow in dense clusters of many small, bright red, rounded pomes (no more than ⅜ inch, or 1 centimeter, in diameter); stick around on tree nearly all winter.

SPECIES REMARKS: The long yet slender thorns, three-lobed leaves, late flowers, and long-lasting fruit set this species apart. Its leaves are similar to Amur maple's (see p. 183) and Golden Raindrops crabapple's (see p. 298), but the maple has oppositely arranged leaves, while the crabapple lacks thorns.

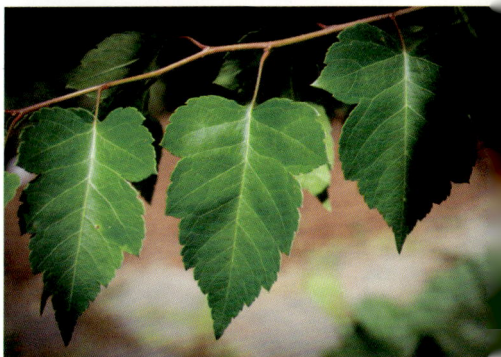

Cockspur Hawthorn

Crataegus crus-galli

This species is planted often as a street tree or in the landscape, mostly for its lovely flower displays and small size, but it also has a somewhat understated architecture in winter, which is what I think maintains its popularity. It's not invasive, so you'll only find it where planted.

Small tree up to 30 feet (9 meters) tall, with more spreading habit; architecture appears angular and interwoven, but not messy, like a bird's nest. **BARK:** Gray on outer layers, but softly and irregularly exfoliating, revealing darker orangish-browns like Lavalle, but stem not spiraled or as seamed. **TWIGS:** Like other hawthorns, with *long, sharp thorns up to 3 inches (8 centimeters) long*; often longer, branched on trunk and older stems; red buds less domed, more oval, with thicker scales. **LEAVES:** 1–4 inches (2.5–10 centimeters) long, *leathery, unlobed with rounded apex*, finely serrated margins. **FLOWERS:** White, borne in flat clusters in mid-spring; emit classic musk of many rose family trees. **FRUIT:** Haws grow in fairly loose clusters; magenta, about ⅜ inch (1 centimeter) in diameter; persists through winter, but not nearly as effectively as Washington.

SPECIES REMARKS: A thornless cultivar is being planted more often, making identification by the thorns less useful. But that absence can be the clue itself, as most others are likely to have obvious thorns, especially older trees. Quite distinctive among hawthorns, the leaves are more rounded and lack lobes.

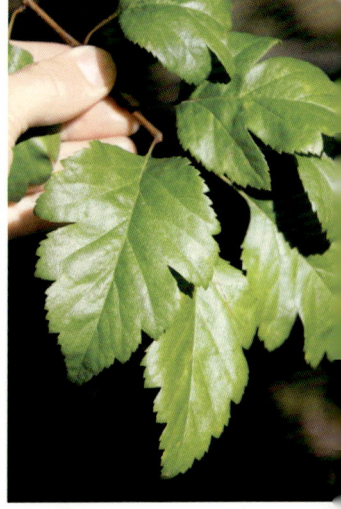

LEFT AND CENTER: *Black hawthorn bark, flowers with leaves* RIGHT: *Winter King green hawthorn leaves*

Similar Species

Black hawthorn (*C. douglasii*) is the region's most common native species. Found mostly in natural areas, and becoming a popular street and yard tree, it is differentiated easily by its dark black fruit and its rounded leaves that tend to get slightly lobed on the upper half near the tip. It blooms very early alongside singleseed hawthorn, but its leaves are larger and without deep lobes, while its bark is intensely flaky and more orange than gray.

Winter King green hawthorn (*C. viridis* 'Winter King') can often be found farther east in our region due to its hardiness. Look for nearly smooth and exfoliating bark, long thorns, and more or less doubly serrated margins on triangular to ovate leaves (see top right photo).

Autumn Glory hawthorn (*Crataegus* 'Autumn Glory') is a cultivar or hybrid that likely has some relation to midland hawthorn; however, it has larger leaves that retain shallow lobes and much larger fruit, even bigger than Lavalle's. The fruit stays on the tree much of winter.

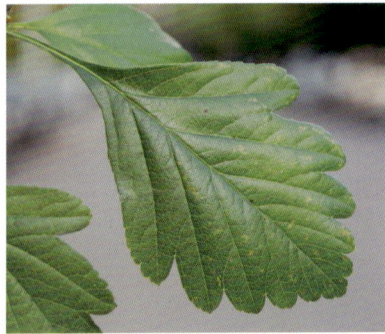

Autumn Glory leaf

SOME HAWTHORN LOOK-ALIKES

Two species that previously were included in the genus *Sorbus* (which is covered later on) have recently been split into new genera. This turned out to be quite convenient for me in the making of this guide because unlike other trees in *Sorbus*, the two below have simple leaves, not compound. I have included them near their closest look-alikes. Sometimes these taxonomic changes have practical benefits.

Whitebeam (*Aria edulis*; syn. *Sorbus aria*) makes an appearance every now and then, with its multistemmed, upright habit and grayish appearance. Its leaves have small, rough-edged lobes or large, rounded double serrations, depending on your point of view. They are glossy green on top but have a white tomentose underside that gives the canopy a light grayish tone. Its flowers are flat-topped corymbs that produce clusters of tiny red pomes in the fall. The bark is smooth and often covered in lichen. Compare this species to **Korean mountain-ash** (*Alniaria alnifolia*; syn. *Sorbus alnifolia*), which has similar leaves except they are nearly glabrous below, along with looser clusters of flowers and fruit.

LEFT SIDE: *Whitebeam flower and leaf underside* RIGHT SIDE: *Korean mountain-ash leaf and developing fruit*

CLOCKWISE FROM TOP LEFT: *Callery pear fruit; Callery pear form; Asian pear leaves*

PEARS

Pyrus

The Pacific Northwest's pears, like the apples, fall into two camps: ornamental and edible, and never the twain shall meet—well, mostly. Of course, the ornamental pears are planted for their flower displays, and they usually produce either no fruit or small, pungent, astringent fruit not fit to be eaten by humans (though other animals seem to quite enjoy them). The edible species also have stunning spring displays, but they produce the same fruit that you would buy at a fruit stand or your local market. Both are quite commonly found in the region.

Enterprising home orchardists often plant edible species and varieties in their yards and planter strips, in the hopes that they'll reap the harvests for years to come. All too often, though, they are left unmaintained and grow out of hand. Multiple varieties are often grafted onto a single stem, such that each main branch could produce unique fruit. But trust me, this is not as confusing as it sounds (see p. 292).

The ornamental species can cause a bit more confusion because their fruit does not have the classic "pear" shape, rather appearing more like small, round apples (which incidentally are what we call crabapples, so maybe these should be crabpears?). But again, there are some good traits you can use to tell the two apart. You'll almost always find pears growing along streets or as yard trees; the most common one, Callery pear, is monstrously invasive in other parts of North America, but it hasn't escaped in our region—yet.

Callery Pear
Pyrus calleryana

Also known as Bradford pear, this is probably the most hated tree planted in Pacific Northwest landscapes, yet somehow it's still one of the most prevalent. I think it's because it promises to have beautiful spring blossoms (and it's dirt cheap). Regardless, it's extremely likely this is the flowering pear you see on basically every other street.

Small tree, usually not over 30 feet (9 meters) tall but sometimes taller; form is usually pyramidal to oval; replete with vertical sprouts or errant limbs arching out of canopy. **BARK:** Gray, rough, broken into vertically oriented, flat-topped, platy ridges with fissures between. **TWIGS:** Stout, brittle, greenish to gray, with lenticels; pubescent only below terminal bud; buds large, conical, brownish, with imbricate scales covered in dense white pubescence; often many short spur shoots. **LEAVES:** Up to 3 inches (8 centimeters) long, ovate or triangular in normal species, more rounded on 'Bradford' cultivar; acuminate tip with crenate margins; *petiole nearly as long as leaf blade*. **FLOWERS:** Perfect, up to ¾ inch (2 centimeters) across, white, with 5 rounded petals; borne in compact corymbs up to 3 inches (8 centimeters) across on spur shoots in spring; famously bad-smelling, musty (look it up for other R-rated descriptions). **FRUIT:** Small, round pome, usually no larger than ½ inch (1.3 centimeters) in diameter; greenish-gold with many lighter dots (see p. 289).

SPECIES REMARKS: There are many cultivars of Callery pear that you can learn about in depth from other books, but the leaves, bark, and twigs will give it away no matter the cultivar or the time of year. Do not plant this tree.

Common Pear

Pyrus communis

Also known as European pear, this is the classic fruit most people think of when they think "pear." The shape, size, flavor, and texture: it's basically self-defining. As the common domestic species, it's grown often in yards and orchards for the fruit, but all too often it's left to grow into a tall, wild mess.

Small tree, often no more than 30 feet (9 meters) tall, with rounded or oval habit (not quite as upright as Callery pear); prone to sprouts, often messy and twiggy. **BARK:** Like other pears, rough, gray, with flattened plates broken up by vertical fissures. **TWIGS:** Stout, yellowish-gray, usually hairless or very slightly hairy; terminal buds imbricate and sharply pointed; plenty of short spur shoots that produce flowers and fruit. **LEAVES:** Up to 4 inches (10 centimeters) long, ovate to elliptic (not triangular as Callery pear), with an acuminate tip; margins entire or dully crenate to serrulate; *hairless and lustrous green*; petiole also very long. **FLOWERS:** Like Callery pear but larger, up to 1.5 inches (4 centimeters) in diameter, set in looser clusters; also smell bad. **FRUIT:** The world-famous pear; ovate pome with grittier cells in the flesh than apples, giving them their distinctive texture; can be yellow, green, or red.

SPECIES REMARKS: The gray, hard, platy bark sets common pear apart from apples or other similar non-pears. Its glabrous, lustrous leaves also set it apart from domestic apple, which has duller leaves with more pronounced serrations along the margin and pubescence below and along the twigs.

MULTIPLE FRUIT VARIETIES ON ONE TREE, YOU SAY?

Pears, as well as a few other fruit trees (namely, apples), are commonly sold and planted as **espaliers**, a very managed form characterized by a single straight stem and several equally spaced horizontal branches all aligned in one plane. This ancient practice was used to grow trees in compact spaces, usually against a wall, which also acted to warm the tree in spring and fall. Today, multiple varieties of one fruit

are often grafted onto one stem, each growing as its own limb (and subsequent tier if grown as an espaliered tree).

Grafting is a propagation technique where a twig (called a **scion**) or a bud from a desirable variety is attached to another, different individual (called the **stock**). If the **graft** survives, then the two parts join and grow as one. Genetically, the two parts remain distinct, but visually, if done well, they appear as one tree.

Usually different varieties of a single species are grafted together, but you may find different species grafted together too. For pears, you're likely to see several varieties of common pear as well as varieties of **Asian pear** (*Pyrus pyrifolia*) on the same

tree. Asian pear is also planted as its own tree and can be identified by the nearly perfectly spherical, baseball-sized, yellowy-russet fruit, which is uniformly speckled with tiny light dots. Its leaves are also longer and skinnier than those of other species, with very sharply serrated margins.

You may also run into a small tree with giant fruit that looks like an oversized, fuzzy, lumpy pear—almost like an apple morphing into a pear—that smells amazing but is as hard as a rock. This is a **quince** (*Cydonia oblonga*). Though it's sometimes planted by itself, quince is commonly used as rootstock for pear trees, and if the pear that was grafted on top dies, or the rootstock is nicked by a weed trimmer, then it's likely to sprout and grow into its own very small tree (up to 15 feet, or 5 meters, tall). The leaves look like a composite of common pear and domestic apple, but with smooth, entire margins and pubescence on all sides. The twigs are also quite pubescent, with large, conical red buds. The main stems are intensely seamed and rippled with a two-toned pattern of orange-browns and armor grays due to a patchwork of exfoliated plates. The flowers are solitary, pinkish white, and easily 2 inches (5 centimeters) wide. Honestly, it's a rad little tree.

Ussurian Pear

Pyrus ussuriensis

Also called Chinese or Manchurian pear, this tree is favored in the coldest areas of the region, namely the far north and east, due to its hardiness and lovely flower displays in spring. You'll find it most often as a street tree but also in yards and parks. It's not reported as being invasive, and it's uncommon in the milder areas west of the Cascades.

Small tree, not much larger than 30 feet (9 meters) tall, with rounded habit; crown usually well-behaved, attractive. **BARK:** Similar to other pears, gray with flattened plates, but with more pronounced furrows. **TWIGS:** Stout, yellowish-brown turning gray, hairless with pronounced lenticels; buds conical, pointed, brown to blackish, usually smaller than other pears; many spur shoots along twigs. **LEAVES:** Up to 4 inches (10 centimeters) long, hairless, ovate to triangular with pointed tip, rounded base; margins finely serrated; petiole very long, nearly same length as blade. **FLOWERS:** Similar to domestic pear; very large, with 5 white petals; borne in clusters along with leaf emergence. **FRUIT:** Rounded pome, fairly large, able to grow to around 2 inches (5 centimeters) in diameter; greenish-yellow with lighter speckles.

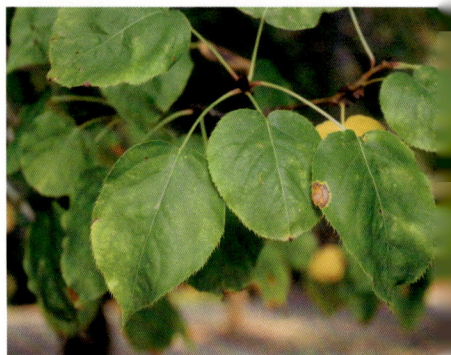

SPECIES REMARKS: In the colder areas of the Pacific Northwest, you're likely to see this species planted along streets. If you're skeptical whether you're looking at one, the darker buds with few or no hairs will set it apart, along with the larger flowers and fruit.

CLOCKWISE FROM TOP LEFT: *White crabapple flowers; classic crabapple bark; flowering crabapples in spring*

APPLES

Malus

Apples are closely related to pears; they resemble each other in several ways, including in their flowers, general form, and large, round pome fruit. Despite their close relationship, you can focus on a few traits that consistently separate them to keep them straight in the landscape. Unfortunately, that's about as far as I'll go because, unlike the pears, apples have been horticultural darlings for centuries—the taxonomy of the ornamental and edible varieties is staggeringly complex.

There are literally hundreds of cultivars of the ornamental crabapples in the landscape and the culinary apples on our tables. Botanically, a "crabapple" and a traditional "apple" are essentially the same thing. The functional definition for the horticulture trade is that crabapples are generally sour and less than 2 inches (5 centimeters) in diameter, while anything larger is a normal apple. Culinarily, though, this distinction breaks down because some crabapples taste great and are used for things like cider—à la Johnny Appleseed, who famously planted all his apples for making frontier cider—and some "normal" apples are horribly sour and mealy. So, it's really an arbitrary distinction.

While the offspring of most trees grow true to form—meaning they look similar to their parent trees—apples do not grow true from seed. Any given seed from any given apple tree may produce a new individual that is wildly different from its parents in almost any way you can think of. This trait (called **extreme heterozygosity**) creates many individuals with different forms, flower types, fruit types, and disease resistance. These new trees with new traits can then be mixed by hybridization to create any number of cultivars with any combination of traits you can imagine—a true breeder's dream, but the bane of the taxonomist.

The crabapples you will find in the Northwest landscape are derived from more than twelve different species and their hybrids; this is roughly the same for the zillions of edible apple varieties as well. I leave it to other authors to explore those complex worlds; instead, I focus on how to tell if it's an apple or a crabapple rather than another rose family tree, and give some examples of a few very commonly found cultivars.

Domestic Apple
Malus domestica

It's the apple that you're thinking of, the one offered by the snake that precipitated the fall (which incidentally was probably a quince, but don't get me started). Its seeds are so variable that it can be semi-invasive. You also may find it as a remnant of a planted orchard that long ago went feral.

with obvious calyx present on bottom.

SPECIES REMARKS: The domestic apple is different from the crabapple by having larger fruit and larger, unlobed, pubescent leaves. The bark is gray and scaly, not shaggy, flaky, or intensely furrowed, which separates it from look-alikes.

Small tree, usually less than 30 feet (9 meters) tall, but unkept trees can be a bit taller; generally rounded or heavily pruned; feral trees look wild, broken, sprouty. **BARK:** Gray, generally smooth, with sparse lenticels; broken up into small platy scales with age, but not overly rough. **TWIGS:** Stout, bronzy gray, with lenticels and pubescence, especially nearer tips; buds red, covered with white pubescence; terminal buds large, conical; lateral buds much smaller. **LEAVES:** Up to 4 inches (10 centimeters) long, oval, with pointed tip and sharp serrations along margin; much larger overall than pears; *underside, petiole, new stems pubescent.* **FLOWERS:** Large, up to about 1 inch (2.5 centimeters) in diameter, borne in clusters in springtime; white with pinkish tinge, usually smell quite pleasant. **FRUIT:** An apple, usually the size of a baseball, green to red,

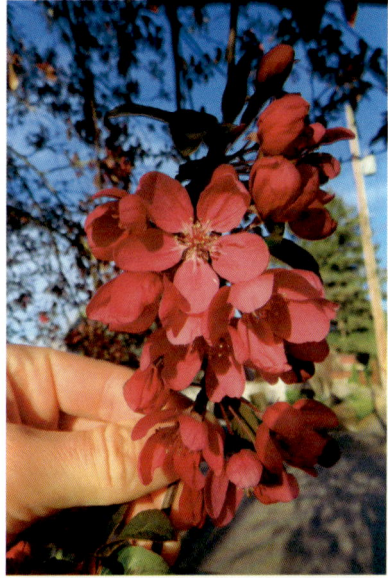

Similar Species

Crabapples (*Malus* [insert catchy name]) are planted as street and landscape trees mostly for their intense spring flower displays and their showy fruit, though sometimes the fruit is simply a necessary tagalong. There are hundreds of cultivars, but what unites them is their profuse flowers; small, rounded fruit (usually less than 1 inch, or 2.5 centimeters, in diameter); and similarly textured bark, which is generally smooth or broken up by long vertical splits or a mosaic of smaller, rougher scales. Everything else, including the leaf shape and color, flower color, fruit color, and form, is highly variable. (See examples above and on p. 298.).

Pillar apple or **Chonosuki crabapple** (*Macromeles tschonoskii*; syn. *Malus tschonoskii*) is a common street tree on the west side of the Cascades. It has a distinctively upright and dense branching habit, and its leaves are coarsely serrated and densely pubescent on both sides at first, then below only. The twigs are very pubescent. Clusters of whitish-yellow flowers beget round pomes, up to 1 inch (2.5 centimeters) in diameter, first yellow, then turning reddish in fall.

CLOCKWISE FROM TOP LEFT: *General crabapple pink and red flowers; pillar apple or Chonosuki crabapple fruit and flowers*

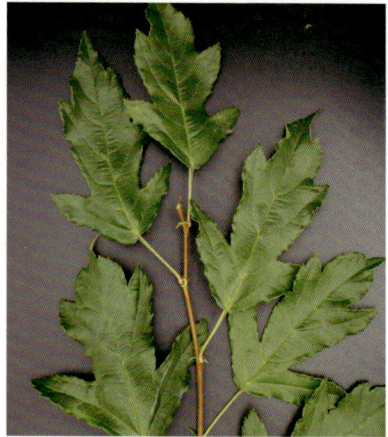

Golden Raindrops crabapple (*Malus* 'Golden Raindrops') is found commonly on the west side of the Cascades. Its rounded canopy is flush with delicate white flowers in mid-spring. The leaves are green and usually lobed (described as cutleaf in some references), with two prominent lobes at the base and one central lobe with serrated edges (similar to Washington hawthorn but with stipules, and Amur maple but alternately arranged, both covered previously). The fruit is a small (1 centimeter in diameter), round, yellow sphere borne in groups on long peduncles. **Royal Raindrops crabapple** (*Malus* Royal Raindrops) is very similar, but with pink flowers, red fruit, and purple leaves.

Prairie Fire crabapple (*Malus* 'Prairifire') is extremely common everywhere because of its large, intensely reddish-pink flowers that cover the tree. It develops a rounded habit with age, but it can have spindly shoots densely covered in flowers when it's younger. The leaves emerge purple, but fade to a dark, red-tinged green over the season, and its fruit is a small, round pome like the others, but it's dark purple.

Then there's Spring Snow crabapple, with white flowers and no fruit, but I'll need to stop there because a section on crabapples could go on forever. *The Tree Book* by Dirr and Warren and the *Manual of Woody Landscape Plants* by Dirr are great resources to dive in for more. Suffice it to say, if you see a profusely blooming tree with relatively large flowers that turn into small, round pomes, and it has smoothish bark, odds are it's a crabapple.

TOP ROW: *Golden Raindrops crabapple fruit and leaves* **BOTTOM ROW:** *Royal Raindrops crabapple fruit and leaves*

PLUMS, CHERRIES, AND THEIR KIN

Prunus

Did you know that cherries and plums are so closely related botanically that they're in the same genus? Well, hold on to your hats because so are peaches, apricots, nectarines, almonds, and the evergreen hedge plants we call English- and Portuguese-laurels.

All these trees have flowers that easily put them in the rose family (Rosaceae) and fruit that puts them in the genus *Prunus*. They differ in small ways that are reminiscent of the pines, a vast genus with a lot of species that all share a common fruit type, but each has its own take on the general form. You can observe these unique takes and narrow down your options fairly quickly with a few purposeful observations.

In the Pacific Northwest, of all the *Prunus* species, cherries, plums, and the two cherry-laurels are by far the most prevalent. Every so often an apricot, peach, or almond will show up, but they're uncommon, so I am focusing on the cherries and plums (see table 20).

Bark: The region's cherries have relatively smooth bark with lenticels; chokecherry and European bird cherry bark is mostly a featureless gray with scant lenticels, and the rest have obvious horizontal lines on smooth to papery bark that ranges from dull gray to shiny purple. Plums have dark brown to blackish bark that is sandpapery rough and furrowed or heavily laden with burls, lacking obvious lenticels. The two cherry-laurels are evergreen, so bark is less important for them.

Leaves: The cherries common to the region usually have larger leaves than those of plums and have more prominent glands at the base of the leaf blade or along the petiole (which plums usually lack). These glands are helpful for identification because other rose family trees lack them too, so you know you at least have a *Prunus* species if you see them. The cherry-laurels look like thicker evergreen versions of cherries, except that they mostly lack glands. None of the region's plums are evergreen.

Flowers: Many common flowering cherries have a small notch at the tip of their flower petals, while plum flowers are oval with no notch. Not all cherries have a notch, but when present, it's a quick giveaway. More depend-able is how the flowers are arranged: cherry flowers are borne either on long racemes (e.g., chokecherry, European bird cherry, Amur cherry, and the cherry-laurels) or in corymbs (e.g., Japanese and other Asian flowering cherries, sweet cherry, and bitter cherry). Either way, there are multiple flowers (and subsequent fruit) collected together. Plum flowers and fruit are borne singly along the twig, though this detail can be hard to see if they are growing quite closely together. Make sure you look at a few twigs to confirm you're seeing the most representative traits. Plums generally bloom nearly two weeks earlier than even the earliest cherries.

TABLE 20. QUICK GUIDE TO CHERRIES AND PLUMS

SPECIES & KEY TRAITS	LEAF	TWIG
Japanese flowering *P. serrulata* Leaves have long acuminate tip with bristles on ends of serrations; flowers in corymbs, often double petaled and pink, blooms later than most others with corymbs		
Sweet cherry *P. avium* Leaves with coarse, blunted serrations; big trees; blooms early with Yoshino; white flowers in corymbs; makes actual cherries		
Yoshino cherry *P. × yedoensis* Early-blooming cherry; leaves with fine serrations, no bristles; famous showy displays in corymbs		
Higan cherry *P. × subhirtella* Earliest bloomer; delicate flowers in corymbs, lanceolate leaves; twigs skinny, often pendulous		

SPECIES & KEY TRAITS	LEAF	TWIG
Sargent cherry *P. sargentii* Usually upright; wide leaves with twisted tip, no bristles; bark shiny purple, buds distinctively pointed outward, blunt; flowers in corymbs		
Amur chokecherry *P. maackii* Mostly found in cold locales; leaves with blunt, minutely wavy serrations; yellow-orange, exfoliating bark and yellow twigs; flowers in small racemes		
Common chokecherry *P. virginiana* Usually purple-leaf variety planted, but escaped in places, so can be green; 2 glands on petioles; flowers in racemes		
European bird cherry *P. padus* Very similar to common chokecherry but usually more glands along petiole, and larger flowers with hairs inside lowest section		

SPECIES & KEY TRAITS	LEAF	TWIG
Bitter cherry *P. emarginata* Rare in built environment; pubescent twigs and small, finely serrated leaves; white flowers in small corymbs		
Cherry plum *P. cerasifera* Early bloomer, full display before most others; flowers borne singly; leaves usually lack glands, papery; makes purple plums		
Common plum *P. domestica* Papery leaves with long tapering base, fine serrations; blooms later than cherry plum; makes blue or yellow plums		

Cherry Plum
Prunus cerasifera

Also called flowering or myrobalan plum, this is the most common plum in our region, welcomed by everyone in early spring as the first truly dazzling display of the year, then often disregarded as a shabby dark cloud of a tree for the rest of it. The purple-leafed varieties are most common, but it has escaped cultivation and can be found along ignored roadsides and the margins of the landscape as a green, thorny thicket.

Small tree up to 25 feet (8 meters) tall, with dense, rounded, twiggy, messy canopy, often with many vertical shoots. **BARK:** Dark brown to blackish, sandpapery rough; stem often spirals, develops furrows (see table 18). **TWIGS:** Brown to nearly black, slender, with many small, round buds, 2 flower buds next to a vegetative bud per node; purple-leafed cultivars generally lack thorns, but escaped trees have long, fairly weak, slender thorns (they still hurt though); buds spherical, tinged pinkish, especially nearing spring (see table 20). **LEAVES:** Up to 2.5 inches (6 centimeters) long, oval with pointed tip and rounded base; fine, regularly spaced serrations along margin; papery texture; petiole sometimes has 2 small glands at base of blade, but more often lacks them.

FLOWERS: Perfect, with 5 petals, usually smaller than 1 inch (2.5 centimeters) wide; numerous, but borne singly along stems, not in clusters or on spur shoots. **FRUIT:** A plum, a more or less 1-inch (2.5-centimeter) long, rounded drupe, *deep red, covered in light waxy bloom*; singly borne, not in clusters.

SPECIES REMARKS: Two main cultivars are most common: 'Thundercloud' has dark purple leaves the whole summer and pink flowers that appear before the leaves. 'Atropurpurea' has more reddish-green foliage when it emerges, and its flowers are much whiter, but still slightly pink with a ruby center.

TOP: *Normal green leaves (left); Thundercloud leaves (right)*
BOTTOM: *Thundercloud (left); Atropurpurea (right)*

303

Similar Species

Blireana plum (*Prunus × blireana*) is commonly found planted as a street tree on the west side of the Cascades. A cross between the purple-leafed cherry plum cultivar 'Atropurpurea' and a double-flowered variety of Japanese flowering apricot (*P. mume*), it has dark purple leaves and dark pink double flowers. It is generally the first plum species to bloom in early spring, just before the cherry plums. It's a deformed-looking tree, though, with its very short stem covered with grotesque warty burls. But the flowers are killer.

European plum (*P. domestica*) is the species we generally eat. It's planted most often as a yard tree for the fruit, but it can sometimes be found as a self-seeded, somewhat invasive thicket lurking along the margins of otherwise polite society. It's a small, shrubby tree with a lot of sprouts, tending to form thickets if left alone too long. Its leaves are green, resembling those of other plums, except that the base tapers much more, giving it an obovate shape. The fruit is mostly dark blue covered in a waxy bloom that rubs off easily, but it may range from yellow to red to blue (see table 18 or photo on facing page).

TOP LEFT: *Blireana plum bark*
TOP RIGHT: *European plum flower*
BOTTOM, CLOCKWISE FROM TOP LEFT: *Blireana plum flower, European plum leaves and twig, Blireana plum leaf*

A NOTE ON THESE "DOMESTIC" NAMES

European plum is in fact a hybrid between cherry plum (*Prunus cerasifera*) and black-thorn or sloe (*P. spinosa*), another European species (that itself may be a hybrid of cherry plum, but with what is unknown). Centuries of breeding have resulted in what is essentially an artificial species, our *P. domestica*. Similarly for other fruit trees, like common pear and domestic apple, this is reflected in their specific epithets, *communis* and *domestica*, respectively.

When botanists and taxonomists started keeping close track of the scientific names of different species, we simply started with what we had and didn't wonder if those were real species or artifacts of our thousands of years of cultivation. Today, we've started to revise some of these names to more accurately reflect the genealogy if we can sort it out. (And that's a big if—remember the weeping willow?) Sometimes, due to quirks of the system, this approach can go comically wrong. For example, the original wild species of our apples is most likely the wild apple of Kazakhstan, once known as *Malus sieversii*. It's currently considered a synonym of *Malus domestica* because if they're the same species, they need to have the same name. Because *Malus domestica* was described first, it takes precedence and thus the wild, original tree is slapped with a specific epithet that means "domesticated", which is a bit of an insult—like calling wolves a variety of domestic dogs.

Japanese Flowering Cherry

Prunus serrulata

The most well-known of the flowering cherries, Japanese flowering cherry (a.k.a. cherry blossom tree) is a very common ornamental species planted throughout the Pacific Northwest. It's planted very often in front yards, in parks, and along the street, but it is not invasive here.

Small tree up to 30 feet (9 meters) tall in cultivation, with rounded to triangularly spreading canopy, often coarse-looking texture; short-lived if not in very agreeable conditions, so often half-dead. **BARK:** Dull gray, smooth but for pronounced horizontal lines of lenticels; sometimes with horizontal peeling strips (see table 18). **TWIGS:** Burgundy to sandy beige depending on whether twig is in sun or shaded; *single lenticels very prominent*; buds conical, pointed, red with several imbricate scales; point away from twig at about 30-degree angle (see table 20). **LEAVES:** Up to 5 inches (13 centimeters) long, more or less oval but with *long, tapering acuminate tip*; emerge red tinged in spring, becoming dark green on top, duller below in summer; glabrous on all sides; margins sharply *serrated with bristly tips* on each serration; usually 2 red glands along petiole. **FLOWERS:** Borne in corymbs, white to pink, single or double petaled, often with wavy or fringed petals; flowers later than other ornamental species with corymbs; often appear as large clumps throughout tree. **FRUIT:** Usually absent (certainly on double-flowered cultivars); if present, a small black drupe.

SPECIES REMARKS: The flowering varieties 'Kwanzan' (also spelled 'Kanzan'), 'Shiro-fugen', and 'Mount Fuji' are the most popular and commonly planted. This species can be easily confused with sweet and Yoshino cherries (following pages) but differs in its larger leaves, more intensely bristled serrations on the margin, and later bloom time (about two weeks).

Sweet Cherry

Prunus avium

Sweet cherry (also Mazzard or bird cherry) is probably the most common cherry on the west side of the Cascades because it's the most invasive. It's also the common edible cherry, producing nearly all the types we eat, but it is also a common rootstock for other ornamental species, often sprouting up if the top dies or is wounded.

Medium tree up to 40 feet (12 meters) tall; largest of region's common species; upright form; unkempt, spreading to arching habit; often growing where it doesn't appear purposeful. **BARK:** Like Japanese flowering cherry, gray, with prominent horizontal lenticels, even on large stems. **TWIGS:** Light brown, turning beige to gray through winter; lenticels obvious, but not as prominent as Japanese flowering's; buds rounded, conical, like eggs with pointed tips; more acute angle between buds and stems than Japanese flowering (see table 20). **LEAVES:** Up to 6 inches (15 centimeters) long, oblong to oval, not as round as others (except for the skinny Higan cherry's), with a wrinkled (rugose) surface and pointed tip; margin coarsely serrated, but *serrations are of varying sizes, all bluntly tipped*; reddish glands at base of leaf blade. **FLOWERS:** White, borne in umbels from spur shoots; 5 petals each, no real notch on tip; appear earlier than Japanese flowering cherry's, but just after others. **FRUIT:** Standard industrial cherry, usually about 1 inch (2.5 centimeters) long, yellow, red, or black (wild, invasive tree's is usually bright red).

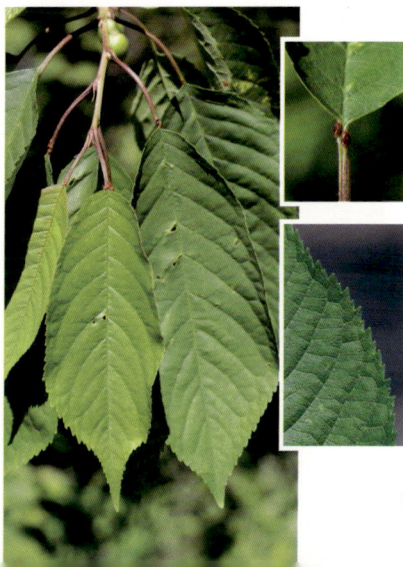

SPECIES REMARKS: Sweet cherry seeds itself in across the built landscape and in unmaintained natural areas nearby, but it is also planted (or maintained wherever it's already growing) for its fruit. The best way I've found to nail it down is the bluntly serrated leaves and the large, white-flowered canopy that lacks a graft union. It also grows large, so if you see a huge cherry, it's very likely this species.

THE CONFUSING CHERRIES

Cherries are not all the same, it turns out, but they can be very difficult to tell apart. In contrast to the shocking amount of variability in the crabapples though, the cherries are much more modest, with fewer species and hybridizations.

Identifying a tree as a cherry is straightforward if you know what to look for. The bark is recognizable for having horizontal lines of lenticels on the main stems. Even on species that develop cracks and furrows, you can still see the remains of these lines on the outer, older plates and scales. The only other trees with similar bark (e.g., Japanese tree lilac and fringetrees) differ by way of their oppositely arranged buds and lack of spur shoots.

Cherry flowers are the next best clue for picking them out from our local lookalikes (namely, the plums, pears, and apples). The flowers of our region's cherries like to grow together, either in elongated racemes with a central peduncle and short-stemmed flowers spirally arranged along it, or in corymbs with all the individual flower stems seemingly arising from a single point. Plum flowers are mostly loners, with individual flowers attached separately to the stems, and while pears and apples also grow in corymbs, they lack cherry's telltale bark. Finally, most regional flowering cherries have very small notches at the ends of their petals, while other species will not.

If you still aren't convinced, look for the tiny little glands resembling little warts present along the petiole or at the base of the leaves. Cherries are the only rose family trees with those glands, which should seal the deal.

CLOCKWISE FROM LEFT: *The double flowers of a Kwanzan Japanese flowering cherry with distinctive notched petals; glands along the petioles of a sweet cherry; spur shoots on a cherry twig*

Yoshino Cherry

Prunus × yedoensis

This is also a flowering cherry from Japan, but it's the species planted in great numbers along waterfronts and plazas that draws huge crowds in the early spring. It's often lumped together with a few other species and generically called a "flowering cherry." It's planted all over, but is not invasive.

Small tree up to 25 feet (8 meters) tall, with usually spreading to rounded crown (see photo, p. 14). **BARK:** Like other cherries, gray, with horizontal lenticels. **TWIGS:** Light orange-brown to gray, with prominent lenticels; buds smaller than others (except for Higan cherry's), tending to point toward end of twig, not as much outward, though there's some variability (see table 20). **LEAVES:** Up to 5 inches (13 centimeters) long, oval or very slightly obovate, with a pointed tip; *emerge green*; glabrous on top, *duller green, slightly pubescent along midrib and veins below*; margin sharply serrated to somewhat doubly serrated, sharper than sweet cherry's but lacking the bristles of Japanese flowering; petiole has slight reddish pubescence and *large, green glands*. **FLOWERS:** Single flowered (not more than 5 petals), white in bud and when open (pink in bud for Akebono), slightly fragrant, profuse, appearing before leaves in early spring; grouped together in corymbs with 4 or more flowers; shallow notch at tip of each petal. **FRUIT:** Occasional; a small cherry, red turning black.

SPECIES REMARKS: Look for the leaf serrations and slight pubescence on the underside to separate Yoshino cherry from other species. It blooms at about the same time as Higan cherry, but its flowers look less delicate due to its more rounded petals. 'Akebono' is a common cultivar.

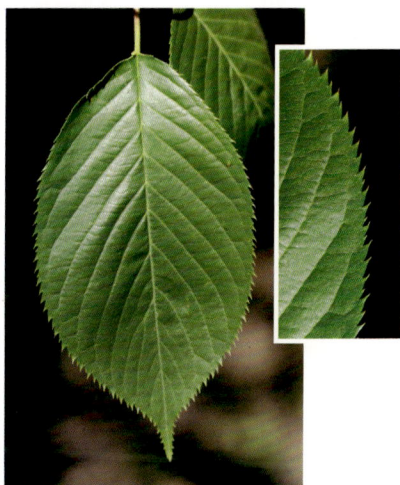

LEFT: *'Akebono' flower* **RIGHT:** *Yoshino cherry flower*

Higan Cherry
Prunus × subhirtella
Syn. *Prunus pendula, P. subhirtella*

Another heavy hitter in the world of Japanese flowering cherries, Higan cherry is one of the earliest bloomers and gets spring rolling. It's been combined with *P. pendula*, which is now listed as *P. × subhirtella* var. *pendula*. You can often find it and var. *autumnalis* planted in the landscape, but neither appears to be invasive in the Pacific Northwest.

Small tree, rarely over 30 feet (9 meters) tall in the Pacific Northwest; rounded to drooping habit with sparse, twiggy canopy; often looks unwell. **BARK:** Like other cherries, gray, with prominent lenticels, but lenticels slightly less obviously horizontal. **TWIGS:** *Slender*, much more so than other ornamental cherries; light brown to olive green with regularly spaced lenticels; often straight and pendulous; buds smallest of ornamental cherries, red, pointed, mostly appressed to twig or at least pointing down toward tip of twig (see table 20). **LEAVES:** Up to 4 inches (10 centimeters) long, smallest of common ornamental cherries; ovate to nearly lanceolate, with acuminate tip and serrated margins; lighter green, more papery texture than others, with slight pubescence on veins below; *pubescent petiole with small, yellow glands*. **FLOWERS:** In corymbs, 2–5 flowers per cluster, often pink or pinkish-white, with distinctly notched petals; sporadically blooms in late winter and warm falls, but full display in spring. **FRUIT:** Occasional, similar to Yoshino cherry's.

SPECIES REMARKS: Many parts of this tree are distinctive; look for the smaller leaves, a twiggy canopy with small buds pointing down the stem, and the thinner, deeply notched petals. If you see a large weeping cherry, it's usually this species (var. *pendula* grafted on top of an upright stem because it can't grow upward). Accolade is a common cultivar, as well as Whitcomb (see below).

LEFT: *Whitcomb flower* **RIGHT:** *Pendula flower*

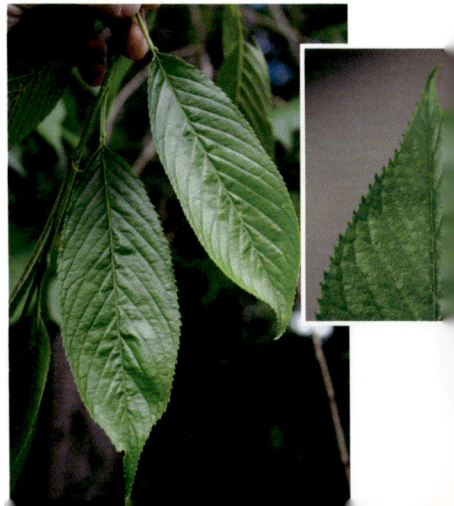

WATCH OUT FOR THOSE GRAFTS!

As noted previously, in the pears section (see p. 292), grafts are common on many ornamental and orchard trees. In fact, most named cultivars of any tree species are clones of the first tree that was found or bred with those traits, and these are propagated continuously from cuttings placed on grafts. The evidence of these grafts is uniquely obvious on cherries, to the point that they can often be dependable identification characteristics.

Graft unions on most cherries appear as strange bulges, usually at the base of a tree or around 4 or 5 feet above it; they are characterized by a change in the size of the stem and in the texture of the bark above and below. These characteristics are particularly prominent when either the graft was poorly executed or the rootstock grows faster than the variety on top, so the lower stem gets wider and develops mature bark more quickly. Sometimes it's both.

It's not uncommon for the grafted cultivar (the top part of the tree) to die back, or simply die altogether, because the genes that helped produce the scion's valuable traits leave the tree less fit in the face of disease and other stressors. When this happens, the tougher rootstock often reacts the only way it knows how: by sending up distress sprouts to keep itself alive. The sprouts in turn grow into stems and end up taking

The graft point on this cherry is easy to spot because the bark changes so dramatically. Grafts often have these stark pattern changes and obvious differences in stem diameter.

over the aerial growing space the top (now dead) used to occupy. The resulting tree is usually a multistemmed mess that doesn't look like the original tree at all. Sometimes it's even an entirely different species.

Sargent Cherry
Prunus sargentii

One of the hardier of our common ornamental species, Sargent cherry is found across the region planted in similar places to the other flowering cherries, though it doesn't seem to be quite as common on the west side of the Cascades as the others. It's not invasive, so you won't find it in natural areas.

Small tree up to about 30 feet (9 meters) tall; most often with upright, vase-shaped canopy. **BARK:** Shiny purplish-gray to brown, with prominent horizontal lenticels; stands out from most others. **TWIGS:** Stout, reddish-brown to beige, with very prominent, yellow-beige lenticels; buds large, reddish, barrel shaped (elongated oval with a somewhat blunted tip); *point away from twig at nearly 45-degree angle* (see table 20). **LEAVES:** Up to 5 inches (13 centimeters) long, ovate, usually widest of common species; *tip abruptly acuminate, usually twisted;* margin sharply and coarsely serrated; emerge bronzy; *large red glands on petiole.* **FLOWERS:** In corymbs, 2–6 flowers per cluster, pink with 5 unnotched petals that don't overlap much (in contrast to Yoshino or Japanese flowering cherries); appear just before leaves, later in spring than others. **FRUIT:** Rare, black, bitter cherry.

SPECIES REMARKS: If you see an upright tree with polished, shiny purplish bark, you've probably got Sargent cherry, although birchbark cherry's bark looks similar (see p. 315). Most often, the cultivar 'Columnaris' is planted, which grows tightly upright, as the name suggests.

Amur Chokecherry
Prunus maackii

Tied for the hardiest of the Pacific Northwest cherries, Amur chokecherry can grow as a gorgeous flowering tree in the far northern and eastern portions of the region. It's planted mostly as an ornamental tree along streets and in parks and yards and is not reported as invasive.

Small tree, usually not climbing above 30 feet (9 meters) tall; rounded canopy with attractive, more open, balanced architecture. **BARK:** *Shiny golden to orange-red,* with horizontal lenticels; exfoliates in horizontal strips; very striking. **TWIGS:** Yellow to orange like bark on main stem, sparse lenticels; buds reddish-orange, conical, appressed to stems (except on terminal spur shoot buds; see table 20). **LEAVES:** Up to 4 inches (10 centimeters) long, oval to semi-lanceolate, with acuminate tip; small serrations along margin; wrinkly surface on top with dots (glands), slight pubescence on veins below. **FLOWERS:** White with 5 petals; long stamens, giving them a bristly appearance; *borne on short, erect, 2–3-inch (5–8-centimeter) long racemes* (not corymbs), with several individual flowers; blooms usually in midspring. **FRUIT:** Small, black drupe that appears in August, also borne on racemes.

SPECIES REMARKS: Amur chokecherry is very rare in warmer climates, so don't expect to find this species unless your area experiences very cold winters. The bark will set it apart right off the bat—there's just no other cherries like it. Some birches have similar bark, but they don't have large, showy flowers or black drupes.

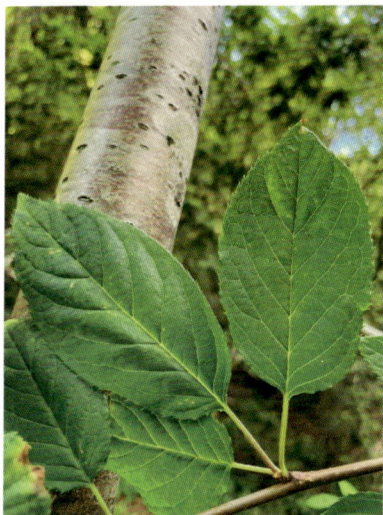

Common Chokecherry
Prunus virginiana

The other super-hardy species, this tree is both a common ornamental planted all over the region and a complicated native species found mostly east of the Cascades, where it grows as a thicket-forming shrub reminiscent of a serviceberry. It's invasive in Alaska, and likely elsewhere, so watch for it in boggy areas.

Small tree up to 30 feet (9 meters) tall, with rounded to oval, often messy-looking canopy; mostly the dark purple variety is planted. **BARK:** Dull gray; lenticels present but not in distinct horizontal lines; rough texture due to lenticels, raised ridges. **TWIGS:** Dark gray to brown with yellow lenticels, glabrous, fairly stout; buds bronzy red, relatively large, conical, with pointed tip; appressed to twigs or pointed slightly away from twigs at acute angle. **LEAVES:** Up to 4 inches (10 centimeters) long, oval to slightly obovate with short, abruptly acuminate tip; minute serrations along margins; usually a purple-leafed variety is planted, which has leaves that emerge dark green, fading to purple by summer; *petiole usually has just 2 glands.* **FLOWERS:** Small, usually no more than ½ inch (1.3 centimeters) across, with 5 white petals; tightly packed on long, 3–6-inch (8–15-centimeter) racemes; appear after leaves emerge. **FRUIT:** Many small red to black drupes that appear in mid- to late summer.

SPECIES REMARKS: Though technically native, only the purple-leafed 'Canada Red' is often found planted. The normal species is rarely found in towns in the Pacific Northwest aside from drainages in drier areas in the east and north, and even then, only as a shrub. The large racemes of tightly spaced flowers immediately set it apart in spring, but look for the smoothish bark and finely serrated margins otherwise.

LEFT: *European bird cherry flower and leaves*
RIGHT, TOP: *Bitter cherry flower*
RIGHT, BOTTOM: *Birchbark cherry bark*

Similar Species

European bird cherry
(*P. padus*), also called Mayday tree, is a European species that has been cultivated there for a very long time; however, it's less common in the Pacific Northwest region than the closely related common chokecherry, which it's similar to in most respects, making them difficult to tell apart. It is invasive in parts of Alaska and Washington, and potentially elsewhere in our region too. The leaves are slightly larger and duller than chokecherry's and are green, not purple or red, in summer. It can have *one to four glands along the petiole* compared to the chokecherry's two, and the flowers are slightly larger, with hairs inside the cupped area at the base (the hypanthium). Finally, the pits in its fruit are rough and wavy in contrast to the chokecherry's smooth pit. The devil is in the details on this one for sure.

Bitter cherry (*P. emarginata*) is the most common treelike native species. Mostly a forest species, it can be found in natural areas or along the margins of semiwild landscapes. The leaves are small, up to 2 inches (5 centimeters) long, and dark green, with a gently pointed tip and very finely serrated margins (see table 20). Its white flowers are borne in small clusters (corymbs) on a slender stem.

Birchbark cherry (*P. serrula*) has extremely shiny bronze to purply bark that *exfoliates in papery strips like a birch.* It has lanceolate leaves and slender twigs like Higan cherry, with a muted flower display. Note that sometimes a stem of a shiny-barked cherry is grafted onto generic rootstock, and then a different flowering variety is grafted on top of it, making for a three-way graft and a very confusing tree.

You may find a few other so-called stone fruit trees out there, like **peach** (*P. persica*), **apricot** (*P. armeniaca*), or **almond** (*P. amygdalus*). All of these are fairly uncommon as ornamental trees, found sparingly in street planter strips or home orchards. Their leaves are all lanceolate and often curved downward, and their flowers are all light to dark pink and rather large (about 1 inch, or 2.5 centimeters, across), but their fruit will quickly set them apart: peaches are big and fuzzy, apricots are smaller and smooth, and almonds are in very fuzzy husks until they split open in the fall. For good measure, a nectarine is just a smooth peach.

315

English-laurel flowers, fruit and leaves

The Two Cherry-Laurels

The last of the common species of *Prunus* are the two cherry-laurel species, both of which are hardly trees most of the time but get up there enough to count. You're probably mostly familiar with English-laurel because it's the most common and most invasive of the two. It's routinely planted as a hedge and left unmaintained, exploding into a huge mess. Portuguese-laurel is found more often as a large shrub or a stand-alone tree. It's not quite as invasive, but that's graded on a scale, as the English-laurel blows it out of the water.

English-laurel (*P. laurocerasus*), also called cherry-laurel, is the large evergreen hedge "tree" with big, thick, light green leaves. If left alone, it grows into a large shrubby tree not much more than 20 feet (6 meters) tall, with big, nearly black-barked stems/limbs. Its leaves are large,

up to 7 inches (18 centimeters), very shiny, and leathery, with nearly entire margins but often with slight bumpy serrations. Its white, five-petaled flowers grow on erect racemes much earlier than flowers grow on Portuguese-laurel. These develop into spherical black drupes that birds love to eat and spread along fence lines. Several cultivars with skinnier leaves and smaller forms are planted.

Portuguese-laurel (*P. lusitanica*) is similar in many respects to English-laurel, but it can grow into a proper tree if left alone. The leaves are slightly smaller, darker green, and not as thick and leathery as English-laurel's. They are also more pointed, with more pronounced serrations along the edges. Its flowers are borne in racemes, but they are longer than the English-laurel's and swoop downward and out. The berries are very similar as well.

Portuguese-laurel flowers, form, and leaves with fruit

Pacific Madrone
Arbutus menziesii

One of the most charismatic and adored native trees in the Pacific Northwest, Pacific madrone, or sometimes madrona, grows on the west side of the Cascades as far north as southern British Columbia. It's most common in drier, rocky natural areas, but it can be found in parks and yards too. It's uncommon as a street tree.
Family: Ericaceae

Large evergreen tree up to 70 feet (21 meters) tall, with upright, but erratic, canopy, almost never growing with a straight, vertical stem. **BARK:** Distinctly smooth, exfoliating when young, with outer reddish-orange bark peeling back irregularly to reveal olive green to beige bark beneath; older stems retain darker brown bark that flakes in vertical strips. **TWIGS:** Stout, light yellow-green when first emerging in spring, turning reddish-brown their second year. **LEAVES:** Evergreen or tardily deciduous, sometimes falling in late winter just before new flush; up to 4 inches (10 centimeters) long, oval, with entire margin; thick, leathery, shiny green on top, duller below. **FLOWERS:** Small, creamy white, urn shaped, with petals fused together with a small opening at end; borne on terminal panicles in late spring. **FRUIT:** Nearly spherical, berrylike drupe with ⅜-inch (1-centimeter) diameter, borne in clusters in fall; orange to red, textured with many tiny bumps.

SPECIES REMARKS: The bark is a quick tell for this species, along with its large evergreen leaves with entire margins. It may be confused with the closely related **strawberry tree** (*A. unedo*, see p. 323).

SPEAKING OF EVERGREEN "TREES"

You may come across a few other evergreen shrubs-but-sometimes-trees that can be confused with one of the cherry-laurels or another common evergreen. Just in case, here's a quick rundown of the most common ones.

Chinese photinia (*Photinia serratifolia*) and **Fraser photinia** (*Photinia* × *fraseri*) are two evergreen shrub-trees also in the rose family that have leaves very similar to those of Portuguese-laurel. Fraser photinia is a hedge plant that can get large but never truly treelike. Its new leaves are bright red when they emerge and often stay that way through the season. Chinese photinia can get to be a small tree (up to 20 feet, or 6 meters), with slightly larger and more leathery leaves and sharp, nearly dentate margins from top to bottom; new leaves emerge yellowish green in spring. The flowers and fruit of both species are small yet borne in big, round-topped corymbs (not long racemes).

Japanese camellia (*Camellia japonica*) and **sasanqua camellia** (*Camellia sasanqua*) are very popular evergreen shrubs (rarely treelike), often planted next to someone's house. Their leaves are thick, leathery, and oval, with pointed tips and serrated margins (each serration tipped with a dark gland). They are in the tea family (Theaceae) along with the stewartias, and accordingly they have similarly large flowers with many yellow stamens. They are hybridized and bred like mad, so they come in multitudes of flower colors and forms. They are simple to tell apart: Japanese blooms in early spring and has hairless twigs, while sasanqua blooms in late fall and has pubescent twigs.

Rhododendrons (*Rhododendron* spp.) are very common shrubs (in the Ericaceae family) that can have a few treelike specimens, but I'll be honest: those are the exceptions. One of the biggest is the Pacific Northwest native western rhododendron (*R. macrophyllum*), but you'll mostly see Asian species or one of the millions of hybrids or cultivars. The easiest way to distinguish them from other evergreens is by their leaves, which spiral around the stem rather than lay in one plane, and by their flowers, which are large and showy, often with their petals fused at the base into a small tube with long stamens and pistils.

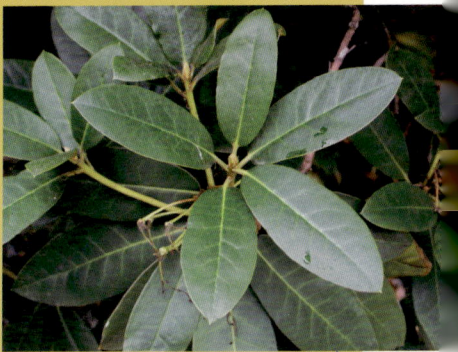

OPPOSITE: *Japanese camellia flower* ABOVE, CLOCKWISE FROM TOP LEFT: *Chinese photinia leaves; Fraser photinia flowers; Japanese camellia leaves; and rhododendron flowers and leaves*

English Holly
Ilex aquifolium

The scourge of gardens and natural areas, English holly is one of the more troublesome invasive species west of the Cascades. It was planted commercially in the past for cut-flower displays and the horticulture trade, but the birds took it from there and it's now a commonly found nuisance. **Family: Aquifoliaceae**

Medium evergreen up to 50 feet (15 meters) tall, with a compact, upright habit; usually one or a few main stems with many skinny, twiggy, and leggy branches. **BARK:** Smooth, gray, sometimes developing horizontal ripples; similar to that of European beech. **TWIGS:** Dark green on newer growth, becoming darker brown, then gray; often arching down, maintaining small diameter back to main stem. **LEAVES:** Evergreen, usually disagreeably pointy; up to 3 inches (8 centimeters) long, dark, lustrous green on top; midvein and lower leaf surface creamy yellow-green; oval with heavily undulating margins replete with extremely sharp and stiff spiny serrations on juvenile foliage; adult foliage often spineless, smooth.

FLOWERS: Dioecious, borne in axillary clusters; up to about ⅜ inch (1 centimeter) across with 4 creamy white petals each. **FRUIT:** Shiny red spherical drupe, usually not more than ⅜ inch (1 centimeter) in diameter, borne in axillary clusters on female trees.

SPECIES REMARKS: You'll know this tree when you walk by it because it'll hurt you, and then you'll despise it too. No other trees in our region have such robust and intensely pokey foliage. Be wary of the variability in the leaves and the lack of berries on the male trees: it's still an English holly and should be treated as such. No other treelike species of holly are common to any degree in the Northwest.

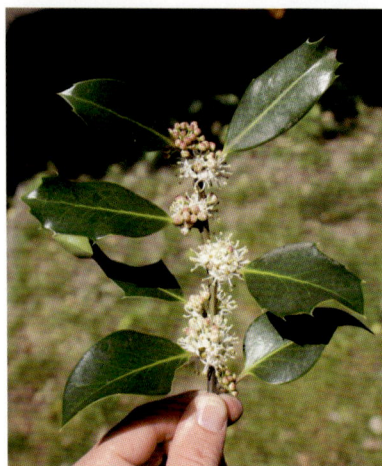

Oregon-Myrtle
Umbellularia californica

Though it grows natively along the southern coast of the Pacific Northwest, Oregon-myrtle (also called California bay-laurel) is planted just about everywhere west of the Cascades. It's not considered invasive, but you'll find it popping up in yards and seminatural areas, sometimes as a street tree or in parks. **Family: Lauraceae**

Large evergreen up to 70 feet (21 meters) tall, with wide, spreading canopy, often large and multistemmed from base. **BARK:** Gray, uniformly patterned, broken into small rectangular sections with shallow fissures between them. **TWIGS:** Slender, green; tend to hold leaves flat and widely spaced in shade, but more radially spread in sun; flower buds globe-like, on stalks borne in leaf axils; look like peas on stalks. **LEAVES:** Evergreen, oval to slightly lanceolate, with a bluntly pointed tip; leathery, somewhat shiny green on top; lighter, duller below; when broken or smashed, they *smell strongly of Vicks VapoRub*; margin entire, *slightly revolute, flat*. **FLOWERS:** Small, creamy white but appear yellow due to stamens; borne in clusters in leaf axils. **FRUIT:** Round, fleshy ball that starts green and fades to brown, held on by a suction cup–like receptacle.

SPECIES REMARKS: This species can be confused with English holly and sweet bay laurel from Europe. Oregon-myrtle will never have sharp spines, and the flowers have more than four petals. Compared to sweet bay laurel, Oregon-myrtle grows far larger, and the leaves are held in a flatter orientation and do not undulate (see p. 323).

321

Russian-Olive
Elaeagnus angustifolia

As English holly is on the west side of the Cascades, so Russian-olive is on the east. It's found growing mostly in riparian areas and disturbed lots, but it is by no means limited to those habitats. Its seeds are dispersed by birds, so it's found wherever they go. **Family: Elaeagnaceae**

Small tree up to 20 feet (6 meters) tall, with rounded, shaggy crown; can be kept as a rounded tree in yards and parks; in wild areas, grows like a thicket, often with angled stems and broken, sprouting branches. **BARK:** Smooth, gray on young stems, soon becoming fibrously shaggy, then rough, flaky; gray to reddish-brown. **TWIGS:** Slender, silvery, usually covered in small scales; often with sharp thorns; older twigs become brown; buds small, rounded to conical, also covered in scales. **LEAVES:** Deciduous, lanceolate to linear, about 3 inches (8 centimeters) long, with entire margins; *dullish green with silvery scales on top*, much more silvery below; look like willow leaves. **FLOWERS:** Perfect, borne singly along stem; silvery in bud, then splitting open with 4 showy, yellow sepals. **FRUIT:** Like a yellow olive dotted with tiny silver scales; technically an achene, but looks like a drupe.

SPECIES REMARKS: You won't get Russian-olive confused with a true olive, because the leaves are alternate (not opposite), covered in scales, and deciduous. True willows won't have silvery scales, and they also have catkins as opposed to single yellow flowers and fleshy fruit.

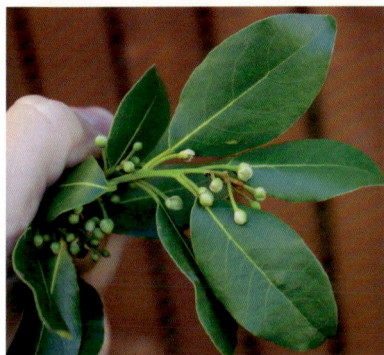

LEFT: *Strawberry tree bark and leaves* **RIGHT:** *Sweet bay laurel bark and leaves*

Similar Species

Strawberry tree (*Arbutus unedo*) is usually a large shrub, but is being planted as a small tree more often. Its evergreen leaves are small (less than 3 inches, or 7.5 centimeters) and dark green with serrated margins. It blooms in the fall when the previous year's fruit matures in bright yellows and reds.

Sweet bay laurel (*Laurus nobilis*), our only true laurel and namesake of the laurel family (Lauraceae), is a somewhat common, small evergreen tree for gardens. Sweet bay laurel (or variably sweet bay or bay laurel) is becoming more common as a small street tree, for better or worse. It's often shrubby and multistemmed, kept small by pruning, as it's the source of the culinary bay leaf. It's found west of the Cascades, as it doesn't prefer very low temperatures. It can get up to 40 feet (12 meters) tall and has an upright, pyramidal form. It can easily be confused with a young Oregon-myrtle, but it has rougher-textured gray bark that does not break into rectangular sections; it also stays much smaller at maturity than Oregon-myrtle. The leaves are more slender and a darker green with *distinctly undulate margins*. Sweet bay laurels are dioecious, so only the female trees will bear their round, olive-like fruit, which is comparatively rare. In contrast, the Oregon-myrtle is monoecious and produces green to light brown fruit all the time.

CLOCKWISE FROM LEFT: *Spinning gum fruit; small-leafed gum form; cider gum leaves*

EUCALYPTUSES

Eucalyptus

Throughout the Pacific Northwest, these exotic trees are called eucalyptus (or shortened to "eucalypt," family Myrtaceae), which is their proper genus name. However, elsewhere in the world they are called gum trees or mountain-ash.

Astonishingly, there are over eight hundred species of eucalyptus, mostly from Australia, and almost all of them are tropical or subtropical trees that don't make it north of the Siskiyous, except perhaps on the far southern Oregon coast. Only those species that grow natively at very high elevations can be planted farther north in the mildest regions west of the Cascades, and even those tend to get hammered by exceptionally cold weather.

Eucalypts have two different leaf forms: Juvenile leaves (those most commonly sold for flower arrangements) are oppositely arranged and conjoined, often encircling the stems (**perfoliate**). Adult leaves, however, are alternately arranged and more lanceolate. Their bark is quite variable, with different patterns ranging from smooth to flaky to rough.

Eucalyptus can be a challenging group because of their variable leaf traits, and I've chosen to only touch on the most common species, though there are a few others that may be out there depending on where you're at. I recommend you visit some local nurseries or botanic gardens that have some eucalyptus species so that you can get acquainted with those that may be in your area.

TOP ROW: *Tasmanian blue gum fruit and leaves* **MIDDLE ROW:** *Bark of Tasmanian blue gum (left) and snow gum (right)* **BOTTOM ROW:** *Snow gum leaves and form*

Along the far southern Oregon coast, you may find **Tasmanian blue gum** (*Eucalyptus globulus*), the giant eucalyptus common farther south in California. It's not found anywhere else in the Pacific Northwest. More often you'll run into a handful of relatively small species, few getting much more than 35 feet (11 meters) tall. No species is especially common, being peddled by just a few specialty nurseries, but this is likely to change as our climate becomes more suitable.

Snow gum (*E. pauciflora*) is probably one of the hardiest and most commonly planted eucalyptuses in the Pacific Northwest. There are two subspecies, *niphophila* and *debeuzevillei*; both have lanceolate, slightly falcate (flattened, laterally curved), blue-gray leaves that are about 4 inches (10 centimeters) long and fairly thick. They have airy crowns, often with a few main stems, and bark that exfoliates in long strips, revealing smooth bark mottled greenish-gray to a creamy beige-white.

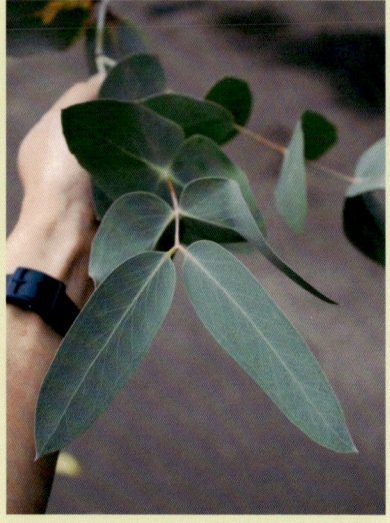

TOP ROW: *Omeo gum bark and leaves* **BOTTOM ROW:** *Cider gum bark and leaves*

Omeo gum (*E. neglecta*) is a medium tree up to about 40 feet (12 meters) tall, with a rounded and denser canopy filled with much wider, more rounded leaves. The juvenile leaves are large and rounded, blue green to nearly purple in winter, and they tend to stick around even on older trees. The adult leaves are likewise wider than others, which helps to set it apart. Finally, its bark is also far rougher and stringier, becoming more furrowed with age, not smoother.

Cider gum (*E. gunnii*) is a larger species, able to reach heights that put some of our other broadleaf trees on notice. Its glaucous juvenile leaves are a brighter green than others and are not always directly connected across the stems, but are instead rather cordate at their base. Adult leaves can be 3 inches (7.5 centimeters) long with a long petiole, shorter, more ovate than snow gum. Its outer gray bark exfoliates in long strips, revealing a brownish gray-green below, also separating it from Omeo.

TOP ROW: *Spinning gum bark and leaves* **BOTTOM ROW:** *Small-leafed gum bark and leaves*

Spinning gum (*E. perriniana*), a smaller tree in the Pacific Northwest, is planted (and named) for its nearly circular juvenile leaves, which look like blue-green discs hanging on a string. The adult foliage is long and thin, up to 5 inches (13 centimeters), tending to get longer as the stem grows farther away from the juvenile leaves. Its bark is a coppery brown that exfoliates in long, thin strips to reveal grayer or greenish bark below.

Small-leafed gum (*E. parvula*)—named for its lanceolate adult leaves that only get about 2.5 inches (6 centimeters) long—grows to a medium height in the Pacific Northwest (see p. 324). Its juvenile foliage is lost very early on, so expect to only see the adult leaves most of the time. In the sun, the twigs are a bright red, which creates a helpful contrast against the gray-green leaves. The exfoliating bark is smooth and gray with patches of light browns, beiges, and olive greens, but is often rough and shaggy near the base.

TABLE 21. QUICK GUIDE TO SMALL-LEAFED BROADLEAF EVERGREENS

While we're covering most of the broadleaf evergreen species, I thought a comparison table might be useful. It includes species with leaves that look similar, but that are spread throughout the book, to help you reference and compare them all easily.

SPECIES & KEY TRAITS	LEAF
English holly *Ilex aquifolium* Very thick and leathery, usually with disagreeably sharp spines (see p. 320)	
Oregon-myrtle *Umbellularia californica* Flat (not undulate) margins, smells strongly of Vicks VapoRub; large tree with square-textured bark (see p. 321)	
Sweet bay laurel *Laurus nobilis* Undulate margins, dark green, smells sweet; small tree with dark, rough bark (not intensely textured, see p. 323)	
Eucalyptus *Eucalyptus* spp. Sharply scented leaves, usually curved (falcate); juvenile leaves round, opposite; thin, peely bark or thick, layered bark (see p. 324)	
European olive *Olea europaea* Lanceolate, gray-green on top, rough to the touch, oppositely arranged (see p. 157)	

SPECIES & KEY TRAITS	LEAF
Silverleaf oak *Quercus hypoleucoides* Margins slightly revolute, entire or with sparse but sharp serrations; dark green above, bright silver and tomentose below (see p. 332)	
Bamboo-leaf oak *Quercus myrsinifolia* Light green on both sides, with slight serrations (almost crenate), somewhat leathery but floppy (see p. 332)	
Cork oak *Quercus suber* Thick, leathery, brittle, with several small, wavy lobes with a tiny, pointed tip; pubescent below; bark corky (see p. 332)	
Canyon live oak *Quercus chrysolepis* Either fully (and sharply) serrated or not at all; thick, brittle, shiny; leaves not as hairy as cork oak's and it has smoother bark	
Pacific madrone *Arbutus menziesii* Leaves large, to 4 in. (10 cm), with entire margins, rounded apex (see p. 317)	
Strawberry tree *Arbutus unedo* Leaves less than 3 in. (7.5 cm), serrated, usually spiraled about stem (see p. 323)	

THE BEECH FAMILY

Perhaps you noticed that I introduced European and American beech way back after the birch family (see p. 220). My reasoning is that beech leaves look far more like hornbeams than the species they're closely related to. I waited to fully introduce the family until now because its main representatives in our region are the oaks and chestnuts. This namesake of the family makes up a relatively small proportion of its members.

Like other families, the beech family is held together by traits of their flowers, which are unisexual. The pollen (male) flowers are produced on long catkins similar to trees in the birch family, Betulaceae, while the pistillate (female) flowers are usually borne singly or in small clusters of just a few. After fertilization, the pistillate flowers mature into fruit that differentiates each genus.

The definitional fruit type of Fagaceae is a nut (also termed a **calybium**) with a cap- or sheath-like covering called a **cupule**. In oaks (and tanoak), this combination manifests as a nut with a relatively small, scaly cap that doesn't usually cover the whole nut and doesn't spit apart (the classic acorn). In beeches, the cupule is larger, semi-spiky, and fully conceals a few nuts until it splits open in the fall. In chestnuts (and golden-chinquapin), the cupule fully encloses a few nuts until it splits open in the fall, but the scales have been transformed into long, stiff spines (often called burs).

CLOCKWISE FROM LEFT: *The long pollen catkin of chestnuts appears on top of the leaves, while on oaks, they hang below; the fruit of a swamp white oak, European chestnut, and European beech.*

LEFT TO RIGHT: *English oak leaves; red oak form*

OAKS

Quercus

Oaks, along with perhaps maples, are some of the most common and recognizable broadleaf trees in the Northwest, despite there being just one species native to most of the western side. The eastern portion has no native oaks, and southern Oregon can claim just three that reach tree size. The bounty of oaks in our region is due mostly to the planting of familiar species from eastern North America and Europe, but more recently, their count has increased as more climate-adapted species from southwestern North America and China are added.

The region's common oaks are separated into two distinct subgenera: red oaks and white oaks. Though there are several botanical and anatomical traits that separate these two, the most important one for you is that red oaks have pointed or bristle-tipped lobes on their leaves, while white oaks have rounded lobes. However, confusion can reign if you happen to find a species with no lobes at all, or an evergreen species that doesn't have the archetypal oak leaf that you may be familiar with.

All oaks make acorns, those little capped nuts we're all familiar with, and it's the definitional fruit of the group. Only a few very closely related species and genera have similar acorn-like fruit, but their flowers differ from the oaks, which sets them apart. Similar to pine cones or maple samaras, acorns are unique to each species. Another broad oak trait, especially for the deciduous species, is that their buds tend to cluster at the tip of the twig, getting set closer and closer together as the twig's growth slows down. These two traits should help get you on the right track; the leaves, form, and specifics of the acorns will get you the rest of the way.

The oak section begins with a quick rundown of the evergreen oaks (and two of their closely related evergreen kin) that are most likely to look like the evergreen species covered earlier, like Oregon-myrtle and English holly. If you're not sure if your species is one of these, check out table 21 on p. 328 to see what's out there.

THE EVERGREEN OAKS AND THEIR KIN

These oaks are the newest additions to the region's pantheon of oaks, but they probably aren't the oaks you'd expect. Evergreen oaks are being planted more on the west side of the Cascades as potential climate-adapted species that will endure our summer droughts, which are expected to become longer and more intense. Though a few are native to the southwest portion of the region, none are especially common outside of their native habitats—yet. Over the next few decades, I expect them to become more prominent members of the region's tree inventories.

Silverleaf oak (*Q. hypoleucoides*) is a small to medium tree from the Southwest. In the Northwest, it stays relatively small, only getting up to 30 feet (9 meters) tall, but it can develop a taller, upright to rounded crown under the right circumstances. Its 4-inch (10-centimeter) leaves are mostly lanceolate, revolute, and entire, but vigorous shoots often produce leaves with large, wavy teeth along the margin. The leaves are dark green and leathery on top but distinctly silver below (see table 21). The acorns are 1 inch (2.5 centimeters) long and light brown, with a cap that covers about a third of the nut.

Bamboo-leaf oak (*Q. myrsinifolia*) is a small-statured tree like silverleaf oak, often confused with it due to its similar unlobed, lanceolate leaves. However, the leaves are a lighter, lustrous green on both sides, and they consistently have minor serrations along the margin and yellow petioles (see table 21). Its ¾-inch (2-centimeter) long acorns are also unique in that the cups have no imbricate scales or bumps, but rather three to six smooth, concentric rings, with a cap that covers about half the nut.

Cork oak (*Q. suber*) is a Mediterranean species well-known for being the producer of all the natural cork in the world, which is in fact its outer bark (and the quickest way to identify it, of course). If the bark doesn't give it away, the leaves are small, about 2 inches (5 centimeters) long, ovate, and very thick and leathery, with a few small, widely spaced teeth along their margins. The top is shiny green, whereas below it's white and pubescent.

Canyon live oak (*Q. chrysolepis*) is native to southern Oregon and getting more attention as a good street tree farther north. It's one of the most variable species of our oaks in both its form (sometimes a shrub; sometimes a tall, arching tree) and its leaves (sometimes entire; sometimes as sharply toothed as holly, often both on a single shoot). Its acorn is usually 1–2 inches (2–5 centimeters) long with a short, dusty yellow cap. If you see a tree that matches this toe to tip but its acorn is longer and pointy, consider looking up **interior live oak** (*Q. wislizeni*).

Two other trees from southern Oregon, the **tanoak** (*Notholithocarpus densiflorus*) and **golden-chinquapin** (*Chrysolepis chrysophylla*), may be planted more often in the future but aren't very common outside their native areas yet. If you find an evergreen tree that reminds you of one of these evergreen oaks or an evergreen chestnut, consider looking into these species.

OPPOSITE, TOP TO BOTTOM ROW: *Cork oak bark and leaves; canyon live oak fruit and leaves; tanoak fruit and leaf; Golden-chinquapin fruit and leaves*

TABLE 22. QUICK GUIDE TO WHITE OAKS

SPECIES & KEY TRAITS	LEAF
Oregon white *Q. garryana* Leaves up to 6 in. (15 cm) long, dark green on top, leathery, often wrinkly; 4 or 5 large, rounded lobes; sinuses can be shallow or deep; most common native	
Eastern white *Q. alba* Leaves like English's but with deeper sinuses; intense purple-red in fall; very uncommon in Pacific Northwest	
English *Q. robur* Leaves up to 6 in. (15 cm) long, green to bluish-green, with 5 or 6 pairs of rounded lobes, shallow sinuses; very short petiole; tiny "earlobes" at base; acorn on long stem	
Two worlds *Q. × bimundorum* Leaves look like a combination of English's and eastern white's, but longer petiole than English and no "earlobes" at base; acorn not as stalked as English's acorn	
Swamp white *Q. bicolor* Leaves up to 7 in. (18 cm) long, light green on top, paler, pubescent below; obovate; many shallow, rounded lobes that can look more like crenate margin near tip	
Bur *Q. macrocarpa* Leaves up to 10 in. (25 cm) long, obovate; rounded lobes bigger, thicker towards tip; few lobes near the base; corky wings on twigs; acorn big, fringed	
Hungarian *Q. frainetto* Leaves 4–7 in. (10–18 cm) long, lustrous green on top, pubescence below; obovate; numerous skinny, rounded lobes, 7 or more on either side; deep, narrow sinuses	

Oregon White Oak
Quercus garryana

Also known as Garry oak, Oregon white oak—the Northwest's most visible native oak—grows throughout the west side of the region to Vancouver Island and just east around the Columbia River Gorge. It's the only native oak north of about Eugene, Oregon. You'll find it natively in dry areas and wet bottomlands and often in yards and parks and along streets.

Large tree up to 90 feet (27 meters) tall, often with massive, nearly spherical crown, whether made of several small trees or one big one; architecture unique in that twigs and stems are curvy and undulate. **BARK:** Gray to light brown, broken up into many small, flat plates or ridges with abrupt fissures; uniformly covering large stems, branches. **TWIGS:** Light brown with lenticels, slight pubescence; buds conical or triangular, with hairy imbricate scales. **LEAVES:** Up to 6 inches (15 centimeters) long, dark green on top, leathery, often wrinkly; 4 or 5 large, rounded lobes on either side of leaf; sinuses can be shallow or deep; some lobes may have 2 rounded tips. **FLOWERS:** Long, pendulous catkins when leaves first emerge. **FRUIT:** 1-inch (2.5-centimeter) long acorn with short stalk or none at all (sessile); cap covers about ¼ of the round, rather plump nut; nut has sharp point at tip.

SPECIES REMARKS: It stands out from afar by its winter silhouette alone due to the rugged, contorted limbs that form a nearly perfectly round outline. The lack of "earlobes" at the leaf base and a longer petiole set it apart from English oak, and the leaves are smaller and darker than other white oaks. **Eastern white oak** (*Q. alba*) looks similar but is far rarer in the region overall (see table 22).

English Oak

Quercus robur

Common on the west side of the Cascades, English oak (also called pedunculate oak) cultivars are planted frequently by people, and not infrequently by squirrels, who quite enjoy the acorns. It's not quite invasive, but lanky volunteers in your garden can be expected.

Large tree up to 60 feet (18 meters) tall, with upright to rounded canopy; most often very columnar cultivar is planted. **BARK:** Hard, gray, broken up into small flat-topped ridges, with relatively shallow furrows between. **TWIGS:** Gray with sparse lenticels; buds red, egg shaped, with symmetrically patterned imbricate scales that *lack any pubescence*. **LEAVES:** Up to 6 inches (15 centimeters) long, green to bluish-green, with around 5 or 6 pairs of shallow rounded lobes; *very short petiole with tiny "earlobes" at base of blade*. **FLOWERS:** In catkins; unremarkable. **FRUIT:** 1-inch (2.5-centimeter)

long acorn with a relatively small cap that covers just about a quarter of the nut; multiple acorns held on *long stalks (peduncles)*.

SPECIES REMARKS: Look for the long stalks on the acorns, the short petioles, and the "earlobes" on the base of the leaf blade to set this species apart quickly. The cultivar Skyrocket ('Fastigiata') is intensely pyramidal. A hybrid between this species and eastern white oak is called **two worlds oak** (*Q.* × *bimundorum*), and a cultivar of it called Crimson Spire is often planted. It has a pyramidal form and the leaves share traits of both parents (see table 22).

Swamp White Oak
Quercus bicolor

A native of eastern North America, swamp white oak has been planted in the region mostly as a street and park tree. It's less common than some other white oaks, but it's becoming more common by the day. It's not invasive, so any naturally seeded specimen is likely a different species.

Large tree up to 75 feet (23 meters) tall, developing a rounded, spreading canopy; often densely packed with leaves. **BARK:** Hard, light gray, becoming flaky with age while developing large, flat-topped ridges with deep furrows. **TWIGS:** Light brown to slightly reddish, stout; buds conical but with rounded tips; *filament-like stipules often present around terminal bud.* **LEAVES:** Up to 7 inches (18 centimeters) long, light green on top, *paler and pubescent below*; wider near the tip, with many shallow, rounded lobes that can sometimes look more like a crenate margin near tip; sinuses variable in depth (compare leaf lobes below). **FLOWERS:** Unremarkable, borne in long catkins in spring. **FRUIT:** Round acorn around 1 inch (2.5 centimeters) long, with a large, bumpy cap that covers ⅓ to ½ of the nut; borne singly or in pairs on long, 2-inch (5-centimeter) peduncles.

SPECIES REMARKS: The wide, usually shallowly lobed leaves with pubescent undersides set swamp white oak apart in summer. Others in the white oak group will have deeper sinuses and smaller acorns with shallower caps. It can be confused with bur oak due to its rough bark, but the filaments around the buds and the lack of corky wings on the twigs set it apart.

LEFT COLUMN: *Bur oak leaves and twigs*
RIGHT COLUMN: *Hungarian oak leaves, acorn, and twig*

Similar Species

Bur oak (*Q. macrocarpa*) is a large (up to 80 feet, or 24 meters), wide-spreading tree that tends to get wider than it is tall. It quickly develops thick, dark gray to brown bark that becomes deeply furrowed and ridged with age; even its twigs tend to develop corky ridges. The leaves are large, up to 10 inches (25 centimeters) long, with an overall obovate shape, and the rounded lobes get bigger and thicker as you go down the leaf. They are pubescent below. The acorns are nearly 2 inches (5 centimeters) long with distinctly fringed cups that cover up to half of the nut. It is relatively common, especially in the eastern portion of the region.

Hungarian oak (*Q. frainetto*) is another large tree (up to 80 feet, or 24 meters), but it tends to remain taller than it is wide, with an upright canopy that is very pyramidal in youth. Its leaves are long (4–7 inches, or 10–18 centimeters) and are lustrous green on top with slight pubescence below. They are also obovate, getting wider as you go toward the tip, but their rounded lobes are numerous and skinny—there are seven or more on either side, separated by deep, narrow sinuses. Hungarian oak's buds are large and hairy, along with the twigs, and its acorns (when present) appear in groups of two to five, each with a bumpy cap that covers about a third of the nut. The most commonly planted variety is Forest Green.

TABLE 23. QUICK GUIDE TO RED OAKS

SPECIES & KEY TRAITS	LEAF
Northern red *Q. rubra* Leaves up to 8 in. (20 cm) long, the broadest of our red oaks, with 7–11 sharply pointed, bristle-tipped lobes; shallow sinuses; fat acorn, tiny cap	
Scarlet *Q. coccinea* Leaves up to 6 in. (15 cm) long, with 7–9 skinny, bristle-tipped lobes; deep C-shaped sinuses; leaf base nearly flat; intensely red in the fall; big acorn cap on wide acorn	
Pin *Q. palustris* Leaves up to 6 in. (15 cm) long, with 5–9 skinny, bristle-tipped lobes; deep sinuses; leaf base is V shaped, sinuses create a U shape; tiny acorn	
Willow *Q. phellos* Leaves less than 5 in. (2 cm) long, lanceolate, lack lobes and bristles, save for a few tiny points here and there; look like leathery willow leaves; tiny, uncommon acorns	
Sawtooth *Q. acutissima* Leaves up to 6 in. (15 cm) long, lanceolate with bristle tips on each serration; acorn has very frilly cap	

Northern Red Oak

Quercus rubra

I bet this is the most common oak tree (often called simply red oak) in our built landscape—full stop. A very common landscape tree across the continent, it has been planted here for well over a century, often helped along by squirrels. It's not considered invasive, but it will grow by acorn if left alone.

Large tree up to 75 feet (23 meters) tall, with massive, rounded canopy held up by huge, long, ascending scaffold limbs; frequently lacking main central leader. **BARK:** Gray, smooth when young, becoming rough with age; tends to break up into irregular ridges, but rarely develops intensely deep furrows. **TWIGS:** Stout, light brown to reddish, hairless; buds red, conical, pointed, angled away from stem with many hairless, imbricate scales. **LEAVES:** Up to 8 inches (20 centimeters) long; the broadest of the red oak group in our region; 7–11 lobes with sharply pointed bristle tips on each; sinuses extend only ⅓ to ½ the distance to the midrib. **FLOWERS:** Borne in pendulous catkins in spring, not helpful for ID. **FRUIT:** Plump acorn, up to 1 inch (2.5 centimeters) long, approaching round; cap like a little beret on a big head, barely attached to the top.

SPECIES REMARKS: Of the three most common oaks in the red oak group, this one has the broadest and fullest leaves, meaning the shallowest sinuses (see table 23). Look for the plentiful acorns with small, flat caps (not giant caps like on scarlet oak's acorns) and an open canopy with giant, mostly smooth-barked scaffold limbs.

Scarlet Oak
Quercus coccinea

Another popular species native to eastern North America, scarlet oak is a common street tree throughout the Northwest, especially on the west side of the Cascades. It doesn't seed itself in, so unlike northern red and pin oaks, you won't find it in natural areas.

Large tree up to 75 feet (23 meters) tall, similar to northern red oak in overall size, but architecture quite different; generally rounded with one central leader for much of its height, then widely spaced limbs; far cleaner canopy than pin oak. **BARK:** Gray, smooth through much of middle age, becoming rough with age, developing irregular ridges, furrows, and generally darker color. **TWIGS:** Like red oak's twigs in overall appearance, but buds lighter in color, less pointed, and scales near tip have slight pubescence. **LEAVES:** Up to 6 inches (15 centimeters) long, similar to northern red oak's but with 7–9 skinny, bristle-tipped lobes, far deeper sinuses that create a C shape between lobes; *leaf base nearly flat*; leaves intensely red in the fall (others aren't as red). **FLOWERS:** Borne in pendulous catkins in spring, not helpful for ID. **FRUIT:** 1-inch (2.5-centimeter) long acorn with a large cap that covers up to half of the nut, much larger cap than red oak's and far larger acorn than pin oak's.

SPECIES REMARKS: The deep, wide sinuses will set scarlet oak apart from northern red oak, along with the large cap on the acorn. Its leaves and form can look similar to pin oak, though. Compare the flat leaf base to pin oak's angled, curved base, and look for a cleaner, well-spaced (not twiggy) architecture.

Pin Oak
Quercus palustris

Of the big three in the red oak group, pin oak is probably the second most common in the region due to copious landscape planting and its propensity to grow from squirrel-planted acorns. It borders on invasive in some areas, growing along fence lines, freeways, and ignored beds, but it doesn't seem to go much farther.

Large tree up to 75 feet (23 meters) tall, with upright canopy; tends to maintain central leader for much of its height; distinctly dense, twiggy; classically, lower limbs angle down, middle ones out, and upper ones up. **BARK:** Gray, smooth when young, then developing rougher ridges with long, shallow fissures between; tends to remain smoothest compared to northern red and scarlet oaks. **TWIGS:** More slender than others, reddish-brown to greenish, slightly shiny; buds smaller than others, conical, pointed, reddish to light brown. **LEAVES:** Up to 6 inches (15 centimeters) long, with 5–9 skinny, bristle-tipped lobes; deep sinuses; leaf base V shaped (not as flat as scarlet oak's); sinuses create a U shape (compared to scarlet's C-shaped sinuses). **FLOWERS:** Borne in pendulous catkins in spring, not helpful for ID. **FRUIT:** Small acorn, up to ½ inch (1.5 centimeters) long, much smaller than others, with equally tiny cap; vertical striations usually present on nut.

SPECIES REMARKS: The dense, twiggy, messy canopy is a quick tell for pin oak, along with the tiny acorns. It's mostly confused with scarlet oak; compare the leaf shape (bases and sinuses) along with the form and fruit.

LEFT: *Willow oak leaves* RIGHT: *Sawtooth oak leaves and fruit*

Similar Species

Willow oak (*Q. phellos*) is a medium tree, up to 60 feet (18 meters) tall, with a rounded canopy; in winter it looks very similar to scarlet oak, but in summer its leaves betray its identity immediately, as they are lanceolate and unlobed, true to its common name. The leaves have wavy margins and are attached in a spiral pattern, giving the canopy a spiky look. Its twigs are very slender, setting it apart from scarlet oak, and the acorns are very tiny, but for whatever reason, they are rare in our region, seemingly not produced in any significant numbers.

Sawtooth oak (*Q. acutissima*) is a medium tree, up to 60 feet (18 meters) tall, with a rounded, broad canopy. The leaves are about 6 inches (15 centimeters) long, ovate to lanceolate, but wider than willow oak's and with distinct bristles along the margin at the end of each lateral vein, much like a chestnut. Its acorns are similar to the bur oak's, with a cap covered in long bristles.

A few other oaks are fairly common across the region, in varying amounts depending on where you're at. For example, if you're south of Eugene, Oregon, **California black oak** (*Q. kelloggii*) may have snuck into your landscape, but it's not often planted farther north or east. Plenty of uncommon species can also be found, planted as unique specimen trees, but they aren't common enough to be included here. If you come across an unknown oak, remember to look for the unique leaf shapes, and find acorns if you can. With a keen eye, you'll notice both the size and number of the lobes and the depth of the sinuses.

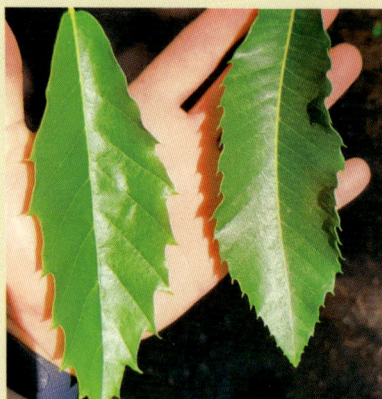

LEFT TO RIGHT: *European chestnut form; comparison of leaves of Chinese chestnut (left) and European chestnut (right)*

CHESTNUTS

Castanea

Chestnuts (family Fagaceae), also called sweet chestnuts to contrast them with horsechestnuts, are common in the landscape, especially west of the Cascades, where you'll find them as stately specimen trees in yards and parks just as often as you will find them as handsome street trees. You're also likely to find a few chestnuts tucked along the margins of the landscape or in unmanaged natural areas, having been planted by industrious squirrels. The best time to see just how common they are is in the fall, when they produce heaps of fruit in intensely spiky husks that litter the ground and betray their hiding spots.

Similar to the elms (covered earlier), the chestnuts are a complicated group to discern with certainty due to their species' similarities and long history of hybridization. The most famous is American chestnut, native to the eastern half of the continent. Today, it's rare to find a mature tree just about anywhere due to chestnut blight, a disease that has all but wiped the species out. Blight has reduced it from the most common forest tree east of the Mississippi to stump sprouts, if you're lucky. American chestnuts planted in the Pacific Northwest were mostly spared from the worst of the blight for a long time, but alas, they are few and far between today.

Most chestnuts found in our region are in fact European chestnuts, or more likely hybrids between several species, including American chestnut, two Asian chestnuts, and European chestnut. Hybrids are the rule today, not the exception. Only the Asian species are relatively resistant to the blight, so they've been bred with American and European species with the goal of boosting resistance while still maintaining desirable traits like size, shape, and tasty nuts. (Similar work has been done to breed American elms resistant to Dutch elm disease.)

In addition to their hybridity, chestnuts are uniquely challenging to identify because of the variability and subtlety of their important characteristics. If your tree has traits that are specific to one species, then you can be relatively confident you've got that species. However, if you keep finding intermediate or a mixture of traits, it's likely a hybrid, and it's perfectly acceptable to declare, "This is a chestnut," and be satisfied.

European Chestnut
Castanea sativa

A common orchard and landscape tree, European chestnut (also called Spanish chestnut) is the most common species found in the Pacific Northwest, and thus the most commonly found self-seeded volunteer along the margins of the landscape.

Large tree commonly up to 60 feet (18 meters); large, rounded to oval canopy, often as tall as it is wide (see photo, p. 344). **BARK:** Gray, rough, with *long, thick ridges* and deep furrows that *characteristically spiral up the stem.* **TWIGS:** Stout, brown, sparsely pubescent if at all; buds plump, rounded with a slight point, hairless; have the shape of a stylized candle flame. **LEAVES:** 5–9 inches (13–23 centimeters) long, oval to lanceolate; sharply serrated margins and *distinctly truncate base; petiole is long,* 1.5 inches (4 centimeters) or more; pubescent when young, mostly glabrous at maturity. **FLOWERS:** Set on long, dense catkins in late spring that shoot out from the twig just above the leaves, often growing along petiole and midvein (see p. 330). **FRUIT:** Round, spiky mass covered thickly in stiff spines; splits in 4 parts, containing 2–3 broadly triangular nuts, often with a small white tassel on their tips; largest nut of common species, over 1.5 inches (4 centimeters) wide.

SPECIES REMARKS: Look for the spiraled bark to set larger, older trees apart. On smaller trees, look for the long, skinny leaves with abruptly flat bases and long petioles, the latter being a particularly good clue that you've got this species or a hybrid with European ancestry.

American Chestnut
Castanea dentata

Though a rarer species, American chestnut can be found both as a remnant of an old orchard or sometimes as a landscape tree in larger, older estates or parks. It can grow from squirrel-planted seeds and often resprouts from the base if the top is killed by chestnut blight, which doesn't kill the roots.

Large tree up to 75 feet (23 meters) tall, with a more upright oval canopy than others; often splits into large scaffold limbs. **BARK:** Gray, developing into flat-topped, interweaving ridges; similar to others, but not discernably spiraled. **TWIGS:** Reddish-brown, more slender than others; glabrous but sparsely covered in lenticels; buds light brown, ovate, with pointed tip; often *angled about 45 degrees from twig* with 2–3 glabrous, imbricate scales; *long, thin stipules near buds*; set slightly off-center from leaf scar below. **LEAVES:** Up to 10 inches (25 centimeters), longest of common species, with short petiole; oblong to lanceolate, with sharp serrations ending in bristly tips; *base distinctly cuneate*, tapering down petiole; apex also tapering to acute tip; *top of leaf dull green* and both sides glabrous. **FLOWERS:** Like other chestnuts. **FRUIT:** Round, spiky mass of spines, but they are shorter, thinner, less intense than others, bending easily; nut similar to others, but small, usually not more than ½ inch (1 centimeter) wide.
SPECIES REMARKS: The leaves are the best way to identify American chestnut: look for their tapering ends—notably the leaf base—and shorter petioles, along with their duller upper surface. The smaller, less stout spines, stipules, and pointed buds help in winter.

CLOCKWISE FROM TOP LEFT: *Japanese chestnut fruit and front and back of leaf*
BELOW: *Chinese chestnut leaf and twig*

Similar species

Japanese chestnut (*C. crenata*) has traits very similar to European chestnut, but the leaves are shorter (3–6 inches, or 8–15 centimeters) and wider, with the widest section closer to the tip. They are also white and glaucous below, and the serrations are no more than bristly points along the margin at the end of the lateral veins. It is the rarest species in our region.

Chinese chestnut (*C. mollissima*) can be found in the landscape, but more commonly you'll find a hybrid between it and the other two described above. Its twigs and buds are pubescent with wide stipules that cover them in summer, and its leaves are shorter and wider, with fewer veins and serrations, and are far glossier than others. The fruit has sharp, stiff spines.

Ginkgo
Ginkgo biloba

One of the most interesting and recognizable trees in our region's landscapes, the ginkgo needs no introduction. It's a curiosity, though, because unlike the flowering plants, or angiosperms, it lacks ovaries, making it a gymnosperm, thus more closely related to the conifers than to the broadleaf trees.
Family: Ginkgoaceae

Medium tree up to about 50 feet (15 meters) tall, with upright but often sparse canopy mostly made up of larger, well-spaced branches; pyramidal when young, but often upright varieties are planted as street trees. **BARK:** Gray with unique pattern of soft ridges and fissures that uniformly cover stems and limbs, gently interweaving up stem. **TWIGS:** Stout, straight to slightly zigzag, reddish-brown; usually many short spur shoots; buds dome shaped, reddish, set directly on newest twigs or atop short spur shoots with many leaf scars below them along the shoot. **LEAVES:** Up to 3 inches (8 centimeters) long, fan shaped, with parallel veins, notch in the middle; held singly on new shoots and in clusters on spurs. **FLOWERS:** Dioecious; technically not flowers, as they do not have an ovary; often called "cones" but appear superficially as catkins; male trees more common, producing pendulous "catkins." **FRUIT:** One big naked seed, nearly spherical, green turning yellow in fall; outer skin smells strongly of vomit as it decays.
SPECIES REMARKS: You'll be able to pick out a ginkgo by its leaves, as there's nothing else like it. In winter, look for the bark pattern and the dome-shaped buds on small spur shoots, or look for its form and texture, which are quite distinctive.

Tuliptree
Liriodendron tulipifera

A common ornamental, also called tulip-poplar or yellow-poplar, it can be found across the region planted as a street tree, a park tree, or every so often as a yard tree. It's a prolific seed producer, but few are viable and none plant themselves, so you won't find it invading anywhere. **Family: Magnoliaceae**

Large tree up to 90 feet (27 meters) tall in the Pacific Northwest, with very upright crown on large, single stem; tends to split into a few main scaffolds, setting it apart from its normal habit in eastern North America. **BARK:** Light gray, smooth when young, but splitting into consistent pattern of interweaving ridges with deep furrows between; very regularly patterned up stem. **TWIGS:** Stout, smooth, light gray to reddish-green, mostly lacking lenticels; *terminal buds paddle shaped with valvate scales* (2 that open like a duck's bill); leaf scars circular. **LEAVES:** Up to 8 inches (20 centimeters) long including petiole, with truncate to indented apex, with 2 pointed lobes on either side and 2 lower down; if you draw whiskers on them, they look like the old cartoon character Felix the Cat; emerge folded on midvein, so have lateral symmetry. **FLOWERS:** Tulip-like, 1–2 inches (2.5–5 centimeters) across, greenish-yellow with orange flares at base of petals; large, elongated stamens. **FRUIT:** Upright collection of winged samara-like achenes; look like ash samaras but with distinct hooklike projection at base; fruit or axis they are attached to often persist through winter.

SPECIES REMARKS: Most of the parts of tuliptrees are quite distinctive, so if you see one trait that seems on point, you can be fairly confident in its identity; if you see two, then there's basically no other option.

349

Sweetgum
Liquidambar styraciflua

One of the most common street trees in the region, and one of the most loathed, sweetgum gives street trees a bad name: it tears up sidewalks. You won't find this tree in natural areas, but it is frequently planted along streets and in parking lots (oh, the irony).
Family: Altingiaceae

Large tree up to 80 feet (24 meters) tall, but often smaller as a street tree; single stemmed with upright canopy, but in open areas upper scaffold limbs grow out irregularly like a mad scientist's hair. **BARK:** Dark gray, rough, quickly developing ridges and furrows that interweave. **TWIGS:** Stout, reddish-green to gray, either rounded or angular with corky wings; buds large, conical, pointed, mostly green to slightly red on ends of scales. **LEAVES:** *Palmately lobed*, with 5, sometimes 7, serrated lobes; up to 8 inches (20 centimeters) long and nearly as wide; very star shaped, lustrous green on top; petiole nearly as long as leaf blade. **FLOWERS:** Borne in early spring as upright, tight clusters, not showy or wildly helpful for ID, but unique. **FRUIT:** Aggregate of capsules forming a woody, spiky ball (like a mace); spikes not that sharp; begin green, maturing to brown in fall; often persist in tree and on the ground around it.

SPECIES REMARKS: Often confused with a maple, sweetgum is easily differentiated by its alternately arranged leaves and buds and its spherical fruit. A variety with rounded instead of pointed lobes (aptly named 'Rotundiloba') can be found, but it isn't very common.

London Planetree

Platanus × hispanica
Syn. Platanus × acerifolia

One of the most common street trees worldwide, London planetree is the result of a hybrid between American sycamore (*P. occidentalis*) and oriental planetree (*P. orientalis*) from eastern Europe. Also called London plane, it's found as a street or a park tree, but never in natural areas. **Family: Platanaceae**

Large tree up to over 80 feet (24 meters) tall in good conditions, with a broad, open canopy; often 1 central leader that splits abruptly into several straight, spreading scaffold limbs. **BARK:** Starts out smooth, creamy yellow-gray, then *exfoliates with age,* leaving a camouflage pattern of yellow, beige, gray; old trees hold on to lower bark, becoming rougher, eventually developing many large bumps, as if stem is slowly melting like wax. **TWIGS:** Stout, yellowish-brown, zigzag at nodes; buds very conical, red to green, with light striations; *concealed by petiole of leaves,* only visible after leaf drop. **LEAVES:** Palmately lobed, with 3–5 primary, triangular lobes, each with distinct teeth along margin and *mostly flat base*; up to 10 inches (25 centimeters) long, blade about same length as width. **FLOWERS:** Pendulous round balls, borne in spring, not wildly helpful for ID. **FRUIT:** Spherical clusters about 1 inch (2.5 centimeters) in diameter without spikes, usually borne in pairs.

SPECIES REMARKS: The bark should set London planetree apart from all but the closely related American sycamore. The large, alternate leaves with broad lobes also help to differentiate it from other palmately lobed trees, and you'll notice that the spherical, spike-less fruit persists through winter.

Bark, leaves, and fruit of American sycamore

Similar Species

American sycamore (*Platanus occidentalis*) is a less commonly planted parent of London planetree, but it's certainly not absent in the region's cities and towns. Planted mostly as a street tree or park tree, it looks very similar to London plane apart from a few telling differences. The leaves of American sycamore have much shallower sinuses and tend to be wider than they are long, creating a more fully filled-in leaf. The base is either flat or more often curved down on either side of the petiole, and it has pubescence along the veins below. Its fruit is very similar, but it is usually borne singly, not in pairs as with London plane. Finally, the young bark is much whiter, and locally the older, lower bark tends to be much rougher and broken into many small plates, some exfoliating off, and lacks bulbous growths.

California sycamore (*P. racemosa*), native to far southern Oregon and California, and **oriental planetree** (*P. orientalis*) from eastern Europe are almost never planted in the Pacific Northwest, but if you see what looks like a sycamore with much deeper sinuses between fingerlike lobes, you might look those two up; the resources listed in the back of the book can help.

Common Fig
Ficus carica

Sometimes a tree, more often an out-of-control, multistemmed monstrosity, this is the only fig that grows in the Pacific Northwest (mostly west of the Cascades) and is a common appearance in home orchards. Though sometimes taken care of, it's more often left to sprout and get way too unruly. **Family: Moraceae**

Small tree up to 30 feet (9 meters) tall, usually multistemmed; creates low, rounded canopy with many limbs near the ground. **BARK:** Starts and remains smooth, gray, with a few splits on older limbs. **TWIGS:** Stout, gray, often appearing jointed with large, circular leaf scars below rounded, almost dome-like buds; terminal buds large, usually very conical. **LEAVES:** Large, up to 8 inches (20 centimeters) long or more on vigorous shoots; usually 3 prominent lobes, sometimes 5, other times none; veins prominent below; margins technically serrated but appear more irregularly entire. **FLOWERS:** Actually figs! The flowers are inside the figs, which are technically very strange inverted receptacles; worth reading about in *The Tree* by Colin Tudge. **FRUIT:** Also a fig; beginning as rough, green balloon; ripening to orange-burgundy, fleshy, slightly ribbed balloon about 1.5 inches (4 centimeters) long. **SPECIES REMARKS:** Figs tend to keep their leaves late into fall, and you'll almost always see the lobed variety. Almost all parts of this tree save for the bark are fairly distinctive, especially when taken together.

MORE TREES WITH VARIABLE LEAVES?

That's right, there's more! Aside from sassafras (see p. 262), mulberries also have variable leaves that tend to randomly have one or two lobes. Only **white mulberry** (*Morus alba*), which is native to eastern Asia, can be found around here. It's a medium tree, up to about 50 feet (15 meters) tall at maturity, but with a fairly spreading crown, and its leaves are large, up to 6 inches (15 centimeters) long, and distinctly smooth and shiny. These leaves can be unlobed with an ovate shape and coarsely serrated margin, but just as often they have one rounded lobe (reminiscent of a mitten) or up to four, two on either side of the midvein. If you can find some blackberry-like fruit, ranging from ghostly white to dark red, you can be sure it's a mulberry. However, the varieties planted in the Pacific Northwest are mostly fruitless, making this challenging. Look for their brownish, shallowly fissured bark with an overall orange tone to confirm.

Bark, leaves, fruit, and twig of white mulberry

As a bonus, **red mulberry** (*M. rubra*), native to eastern North America, looks almost identical except that its leaves are duller, their margins are more sharply serrated, and they're distinctly rough to the touch.

ALTERNATE, COMPOUND, PINNATE

Just as with the trees with oppositely arranged, pinnately compound leaves, trees in this group are quickly differentiated once you find the buds, as they inform the leaf type and arrangement. There are no trees in our region with alternately arranged, palmately compound leaves, so this narrows down species considerably.

Compared to oppositely arranged, pinnately compound leaves, this group is far more varied in its constituent species and genera. Therefore, it's especially important that you pay close attention to species' and genera's unique traits; recall that flowers and fruit are especially important in this regard, and make sure to always look for those first after the leaf arrangement and type. If these reproductive parts aren't available, bark and leaflets are your next best clues.

These trees also have a wide array of bark types, running the gamut from smooth and featureless to rough and thorny; these traits alone will help narrow down your options substantially. The leaflets are similarly varied across the different species, having unique shapes,

arrangements, numbers, and even smells. In particular, notice the presence or absence of a terminal leaflet. Most pinnately compound leaves are **odd-pinnate**, which simply means they have an odd number of **pinnae** (that is, leaflets). This is usually because they have paired leaflets along the rachis, as well as one final terminal leaflet at the tip. An **even-pinnate** leaf lacks the terminal leaflet, so it has an even number of pinnae. This great identification characteristic occurs on only a few species.

This group is dominated by one family: the pea family, Fabaceae. This family has the most variety in leaf type of any family covered in this guide—every type is represented except for palmate. The one type unique to the pea family, and to this section, is the bipinnately compound leaf. Compound leaves that are **bipinnate** have pinnately arranged leaflets that are pinnately compound, creating leaves that remind me of fractals, those artistic patterns that repeat on smaller and smaller scales. Our few trees with leaves like this are sure to catch your eye when you find them.

Tree of Heaven
Ailanthus altissima

Likely one of the most despised invasive species in our region, as well as many others, tree of heaven is well-known for its amazingly fast growth and ability to colonize disturbed areas, ranging from roadsides to cracks in the sidewalk. You'll find it wherever people are, save for higher elevations.

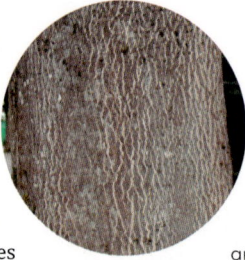

Medium tree up to 60 feet (18 meters) tall, broadly vase-shaped canopy on a single stem when well-behaved, but often clumps of sprouts around larger stem. **BARK:** Smooth, gray when young, developing a gray diamond pattern that morphs into long, interweaving striations. **TWIGS:** Gray to light brown, very stout (especially on new sprouts, often as big or bigger than your finger); curved upward, slightly knobby; buds round, domed, slightly pubescent, directly over large Pac-Man-like leaf scars with an obvious semicircle of vascular bundle scars. **LEAVES:** Huge, not uncommonly 2 feet (60 centimeters) long, sometimes larger; 13–25 leaflets, each 3 inches (8 centimeters) or longer, *usually with at least 1 small lobe at* lower base; *unpleasant nutty odor when crushed.* **FLOWERS:** Dioecious mostly, but variable; both flower types borne in terminal clusters of small, creamy white flowers in early summer; male flowers are malodorous. **FRUIT:** Flat or slightly twisted samara with a single seed in the middle; borne in bunches at ends of branches; reddish-orange, becoming dusty brown; persistent through winter. **SPECIES REMARKS:** Often confused with black walnut, tree of heaven's leaves are much larger, and the leaflets have fewer, less obvious veins and *usually one small lobe.* They also smell far worse: nutty as compared to citrusy.

CLOCKWISE FROM LEFT: *Pure joy finding a massive butternut (with fruit inset); Caucasian wingnut leaves; Caucasion wingnut fruit*

WALNUTS

Juglans

This family of nuts (Juglandaceae, which includes the hickories and wingnuts) is probably one of the most famous and well-known, recognized as much for their tasty fruit as for their useful wood. Most of the representatives in the Pacific Northwest are from eastern North America, but a few from Asia and Europe have made their way here too. They all share large, alternately arranged, pinnately compound leaves and usually large, capacious canopies. They also all technically have nuts for fruit, but the wingnuts' are hardly recognizable as such.

Walnuts are the most common, large, nut-producing trees in our region; hickories, wingnuts, and butternut are all scarcer, which makes it both exciting and confusing when you run across one. All are often mistaken for ash (*Fraxinus* spp.), which also have pinnately compound leaves; however, they are easily differentiated by looking at the buds: ash leaves are oppositely arranged, while all Juglandaceae family members in our region have alternately arranged leaves—another good example of why it's so important to understand what is and what is not a leaf when identifying trees.

The three genera in Juglandaceae in this region can be separated easily enough by looking at their fruit. Wingnuts (specifically, **Caucasian wingnut**, *Pterocarya fraxinifolia*) are the obvious oddballs; they have long, pendulous chains of nutlets no more than ¾ inch (2 centimeters) across, each with two wings on either side, and naked buds. But, as they are entirely uncommon in our region, I'll leave it at that. Walnuts (*Juglans*) and hickories (*Carya*), on the other hand, have hard-shelled nuts surrounded by a husk, the primary difference between them being that hickory husks split apart into four sections, while the walnuts don't split cleanly at all.

Black Walnut
Juglans nigra

A grand and picturesque tree, black walnut is a common sight in parks, along streets, and in backyards, having been planted far and wide for its beautiful hardwood and tasty nut (though the English walnut is preferred as a crop). It is known to grow from squirrel-planted seeds, so you may find it popping up in strange places.

Large tree up to 75 feet (23 meters) tall, with large, rounded, spreading crown; often single stem that splits into a few massive scaffold limbs; rugged winter silhouette resembles a scary, shadow-casting tree in a spooky movie scene. **BARK:** Dark gray, nearly black, rough; broken into irregular plates separated by horizontal cracks and deep vertical fissures. **TWIGS:** Stout, grayish-brown, mostly straight, with pubescence; *pith distinctly chambered like others in genus*; terminal bud ovate to pyramidal, downy; lateral buds globe-like, downy, often in stacked pairs, perched above large, heart-shaped leaf scars. **LEAVES:** Large, often up to 1.5 feet (45 centimeters) long, with 15–23 paired leaflets and single terminal leaflet (sometimes missing); each leaflet 2–5 inches (5–13 centimeters) long, ovate to lanceolate, slightly serrated; *sharp, citrusy odor when crushed.* **FLOWERS:** Monoecious; pollen flowers in tight clusters along pendulous catkin, pistillate flowers small, at ends of twigs. **FRUIT:** Walnut, up to 2 inches (5 centimeters) in diameter; encased in *spherical*, green, lightly textured husk.

SPECIES REMARKS: Distinguish it from ash by its alternately arranged leaves and buds, and from tree of heaven (and all other trees with pinnately compound leaves) by its citrusy-scented leaves; round, golf ball–sized fruit; and dark black bark.

English Walnut
Juglans regia

Probably the most common walnut in our region, this is the well-known crop tree grown in vast orchards here. Also called Persian walnut, it is often found as a street or yard tree. It's frequently planted by squirrels, growing in sometimes inopportune areas of the landscape.

pollen flowers uniformly spaced, not clumped together. **FRUIT:** Walnut; encased in *egg-shaped*, green, lightly textured husk; smallest of our walnuts. **SPECIES REMARKS:** The light gray, ridged bark; smaller leaves with far fewer leaflets; and egg-shaped fruit set this species apart. You may also find **butternut** (*J. cinerea*); its leaves look more like black walnut's, with eleven to nineteen lanceolate leaflets, but its bark and fruit are very similar to English walnut's, though butternut's fruit is bigger and pointed like a lemon (see photos on p. 357).

Medium tree up to 60 feet (18 meters) tall; the upright habit when young develops a large, rounded, spreading canopy; architecture not as ominous as black walnut's. **BARK:** *Light gray*, smooth when young, but with age develops very wide, long, interweaving, *flat-topped ridges* separated by deeper fissures. **TWIGS:** Stout, grayish-brown to reddish, often crooked, bending gently at nodes; pith chambered, brownish; terminal bud triangular to spade shaped, dark gray, pubescent; lateral buds dome-like, often on short projections above curvy, V-shaped leaf scar; flower buds look like erect cones. **LEAVES:** Shorter than others, with just 5–9 leaflets, rarely more, each 2–5 inches (5–13 centimeters) long; *distinctly more rounded and widely spaced than black walnut's;* leaflets get larger as you go toward end of leaf. **FLOWERS:** Similar to black walnut's, except

WALNUT HYBRIDS: YOU THOUGHT IT WAS THAT EASY?

There are two hybrid species of walnut that can be found in the Northwest, both sharing parentage from a little-known species from California and very southern Oregon called Northern California walnut, or simply Hinds walnut (*Juglans hindsii*). This species looks very similar to black walnut (*J. nigra*) in both leaf and bark characteristics, but it has a smaller, rounder nut with thicker, nearly smooth shell walls.

The hybrid between Hinds and English walnut (*J. regia*), called Paradox walnut (*Juglans* 'Paradox'), has been planted as a landscape and timber tree due to its very fast growth and lovely wood, tending to reach massive sizes relatively quickly. The hybrid between Hinds and black walnut is called Royal walnut (*Juglans* 'Royal'). (Neither has a proper hybrid name because the genetics are slightly dubious, but I'll spare you that swamp of confusion.)

Both hybrids can be found sparingly throughout the region as landscape or old orchard trees. Royal walnut appears in the southwest region most often as that's where Hinds walnut is found natively and where eastern black walnut is planted ornamentally. All three are extremely hard to tell apart. Paradox has intermediate leaf and bark characteristics between its parents, but it tends to have eleven to fifteen leaflets and more square, thickly shelled nuts (more like Hinds). Royal tends to have seventeen to nineteen dark green, lanceolate leaflets (but can have between eleven and twenty-three) with sharply serrated margins and a pubescent rachis. Its nuts are also smoother than black walnut's, a trait of its Hinds parentage.

I can only wish you good luck in sorting through these hybrids, but suffice it to say that the two main species covered above are still the most common.

CLOCKWISE FROM LEFT: *Shellbark hickory bark; shellbark hickory fruit; shagbark hickory twig*

HICKORIES

Carya

Hickories are fairly uncommon in the Pacific Northwest, found mostly as street or park trees or, in the case of pecan, as old orchard trees. They tend to be difficult to transplant and don't seed themselves in easily, so despite their attractive canopies, they have never taken off here as popular landscape trees. In eastern North America, where several species are native, they grow at will.

Hickories have alternately arranged, pinnately compound leaves with an odd number of leaflets (odd-pinnate) that look similar to walnuts'. However, their bark does not develop furrows or thick ridges—instead developing thin, crosshatched ridges or large, semi-exfoliating, woody plates—and their nuts are enclosed in a husk that splits apart into usually four distinct sections. Observing these traits will quickly help you differentiate the hickories from other species with pinnately compound leaves, and a few traits about each one will set them apart from each other.

Pecan
Carya illinoinensis

Pecan is famous for its delicious, sweet nuts, so often people are surprised to hear that pecans are a species of hickory—trees that have nuts but are far more well-known for their wood. In the Pacific Northwest, pecans are comparably uncommon, but they can be found as park and street trees or large remnant giants from old orchard plantings.

Large tree up to 75 feet (23 meters) tall, with a large, rounded, bushy canopy; usually a single leader with a few large scaffold limbs. **BARK:** Light gray to light brown, developing thin, flat-topped ridges that interweave up the stem, becoming rough with age; uniformly patterned. **TWIGS:** Stout, gray to light brown, with light pubescence, especially when young; buds large, light brown, plump, somewhat spade shaped with pointed, often slightly hooked tip; pubescent. **LEAVES:** 8 inches (20 centimeters) long or more; odd-pinnate with 9–17 lanceolate, coarsely serrated, *distinctly sickle-shaped (falcate)* leaflets, each on a short stalk (petiolule) and with an *offset base*. **FLOWERS:** Dense clusters of pollen catkins in spring; female flowers near end of twig, inconspicuous. **FRUIT:** Thin-shelled, egg-shaped hickory nut; outer husk green, also thin, with pointed tip and 4 raised ridges when young, like a little blimp; splits into 4 sections when mature, revealing darker brown nut inside.

SPECIES REMARKS: The sickle-shaped leaflets with slightly oblique bases are the best clue for pecans when the leaves are present. If you can find a pecan nut, know they are also unique in that all the other hickories have far thicker husks and smaller nuts that aren't nearly as tasty.

ABOVE: *Shagbark hickory bark, leaf, fruit, and leaf margin hairs*
BELOW: *Shellbark hickory leaf and leaf margin hairs*

Similar Species

Shagbark hickory (*C. ovata*) and **shellbark hickory** (*C. laciniosa*) are two other species that you may find once in a while that can look very similar. Both are large-growing trees that can reach 70 feet (21 meters) tall, with rounded crowns; they both have large buds and stout twigs (see p. 361); and they both have gray bark that starts out very uniformly cross-hatched but develops large, reflexing plates, which give them their names. To tell them apart, look to their leaves. Shagbark hickory usually has just *five obovate leaflets* with yellow pubescence below and *distinct tufts of hairs* along the margin, concentrated at the tips of the serrations; shellbark usually has *seven leaflets* (sometimes five or nine) with pubescence below and along the margin, but it's uniformly dispersed, not concentrated in tufts. Both fruits are round, but shagbark's nut is much smaller than shellbark's, which has a much thicker husk (see p. 361).

THE PEA FAMILY

You've already learned about one genus in this fabulous family (Fabaceae), the redbuds (*Cercis*), but most of the other genera and species represented in the Northwest have pinnately (or bipinnately) compound leaves, so they are grouped together here. The pea family can be picked out easily enough by its fruit, which is the quintessential "pea pod." Pea pods, or legumes, are usually flattened and have two suture lines that allow the pods to break open into two halves like a hot dog bun. Just like the different cones of the pine family, different genera and species in the pea family have different variations on this archetypal form that can quickly help you tell them apart.

Most of the region's common species (silk trees and coffeetrees being the striking exceptions) have the well-known five-petaled "pealike" flower. In contrast to flowers in, say, the rose family, which have radial symmetry, meaning they are symmetrical no matter which way you spin them, pea flowers have bilateral symmetry, meaning they are symmetrical only when split left and right, like a human face. They also share a unique petal arrangement: one up-flipped "banner" petal on top, two "wing" petals below on the left and right, and two fused petals called the "keel" that surround the reproductive parts. Though it seems complicated, this flower form has clearly been wildly successful: Fabaceae has over nineteen thousand species across nearly nine hundred genera.

I mentioned silk tree (or mimosa) as one notable exception to the classic pea flower morphology in our region. Mimosoid flowers are radially symmetric but appear as large puffs of showy stamens, with barely noticeable petals. Over 2,500 species share this flower type, including the mesquites and acacias of the Southwest and elsewhere. The fruit, though, is still the standard legume pod, firmly placing it in the pea family. (Kentucky coffeetree also has radially symmetric, but inconspicuous flowers.)

Take care not to confuse catalpa pods with pea pods, even though one vernacular name for catalpa is string bean tree. Catalpa pods are filled with little, mustachioed, winged seeds, not round, unwinged seeds.

TABLE 24. QUICK GUIDE TO THE PINNATE PEA FAMILY

SPECIES & KEY TRAITS	LEAF
Yellowwood *Cladrastis kentukea* Leaflets alternately arranged down rachis; petiole base covers bud, so leaf scar nearly encircles it on twig	
Amur maackia *Maackia amurensis* 7–11 rounded, glabrous leaflets (11–13 and pubescent for Chinese maackia); bark has peely pattern; tightly packed panicles of small white flowers	
Japanese pagodatree *Styphnolobium japonicum* 7–17 long, pointed leaflets, glaucous below; petiole covers jet-black buds; twigs lime green; pods constricted between seeds	

SPECIES & KEY TRAITS	LEAF
Black locust *Robinia pseudoacacia* Long leaves with a lot of leaflets (7–19); pair of spines at base of petiole; pod small, papery; bark thickly ridged	
Honeylocust *Gleditsia triacanthos* Pinnate on older shoots, bipinnate on new; many small leaflets; platy bark with long, branched thorns; pods longest, at 8–18 in. (20–45 cm), flattened yet often twisted	
Silk tree *Albizia julibrissin* Bipinnate leaves with dozens of small leaflets, very finely textured; flowers are poofs of pink stamens, fragrant; bark smooth and gray; new pods often present with current year's flowers	
Kentucky coffeetree *Gymnocladus dioicus* Huge, bipinnately compound leaves, can be over 3 ft. (1 m) long; pod thick, leathery, the size of your hand; bark is platy like honeylocust, but plates are smaller, giving it a finer textured appearance	
Goldenchain tree *Laburnum × watereri* Trifoliate, small shrub/tree; long chains of yellow flowers in mid-spring; long chains of small, papery pods afterward; twigs green, bark smooth with diamond-shaped marks	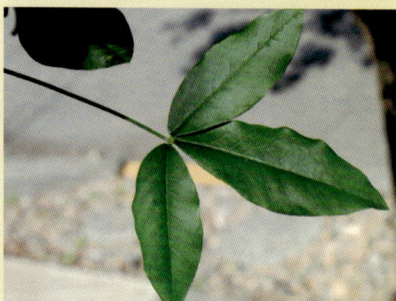

Yellowwood
Cladrastis kentukea

This native of the mountains between North Carolina, Kentucky, and Tennessee is an ornamental beauty planted often as a street tree and landscape tree. Even though it's not invasive, it's one of those trees that you'll suddenly start to see everywhere after you learn about it.

Medium tree up to 50 feet (15 meters) tall, with broad, rounded canopy; splits into several main limbs fairly low, which grow very uniformly up and out. **BARK:** Consistently smooth, light gray, only developing shallow cracks when stems get large; often spotted with white lichen. **TWIGS:** Stout, smooth, without hairs, coppery reddish-brown; buds pyramidal, often pointing away from stems at sharp angle; technically naked, but appear as pubescent cones; *petiole base completely encloses buds*; leaf scars thus nearly encircle buds. **LEAVES:** Large, 8–12 inches (20–30 centimeters) long, odd-pinnate, with 7–9 leaflets; *leaflets distinctly alternately arranged down rachis*; leaflets ovate to slightly obovate (very magnolia-like), light green with entire margins; biggest leaflet at tip of leaf (see table 24). **FLOWERS:** Perfect, white, appearing in late spring in pendulous terminal panicles; fragrant. **FRUIT:** Flattened pod, up to 3 inches (8 centimeters) long, several hanging in pendulous clusters in late summer.

SPECIES REMARKS: With leaves present, the large, alternating leaflets are the best clue to yellowwood, especially if paired with its flowers or fruit. Check that the buds are enclosed by the petiole (or have a ringed leaf scar) to confirm.

Amur Maackia

Maackia amurensis

This tree has gained a bit more prominence lately for being tough enough to be planted along streets and below power lines, given its spreading yet short habit, and I imagine it'll increase in popularity over the coming decades. Not invasive, it is mostly found along streets and in parks.

Small tree up to 30 feet (9 meters) tall, with rounded canopy that spreads out widely with age, losing its central leader. **BARK:** Coppery-green when young, brownish-gray with age; develops unique peeling texture, furrows. **TWIGS:** Shiny gray, glabrous, with diamond-shaped lenticels; buds prominent, smooth, dark brownish-gray, almost plastic-like, with 2 scales; ovoid to slightly conical, plump. **LEAVES:** 8–12 inches (20–30 centimeters) long, odd-pinnate with 7–11 glabrous leaflets; leaflets no more than 3 inches (8 centimeters), becoming more angled toward leaf tip as you go down the leaf. **FLOWERS:** Perfect, white to slightly greenish; many small flowers borne tightly on upright, terminal racemes; often multiple racemes at same point on twig (thus tech-nically a racemose panicle).
FRUIT: Flattened pod, up to 2 inches (5 centimeters) long, held initially outright on racemes, but eventually pendulously; often persists through fall.
SPECIES REMARKS: The dis-tinctive bark should help set it apart quickly, but also note the tight, upright flower clusters and very small fruit. **Chinese maackia** (*M. hupehensis*; syn. *M. chinensis*) is very similar but has eleven to thirteen leaf-lets that are smaller and emerge with pubes-cence, rather than being glabrous.

ABOVE: *Both species have similar flowers and buds* BELOW: *Chinese maackia leaf*

Japanese Pagodatree

Styphnolobium japonicum

Mostly found along streets and sometimes in parks, pagodatree seems to blend into the background until you become aware of it, and then you'll notice it everywhere. It's a very tough tree, but it's not invasive, so you'll only find it where it's been planted.

Medium tree up to 50 feet (15 meters) tall, with rounded but sometimes irregularly spreading canopy. **BARK:** Rough gray-brown that splits apart to create interweaving ridges, giving somewhat braided appearance. **TWIGS:** *Green* for the first few years, basically hairless, but with lenticels and slight roughish texture; buds jet-black, woolly, covered by swollen base of the petiole, appearing set into stem after leaves fall. **LEAVES:** 6–10 inches (15–25 centimeters) long, odd-pinnate with 7–17 leaflets; leaflets up to 2 inches (5 centimeters) long, *oval with pointed tips*; top shiny green; glaucous white below. **FLOWERS:** Perfect, white, borne in large terminal panicles up to 12 inches (30 centimeters) long in mid- to late summer; very showy; whole tree has appearance of fireworks, as each panicle looks like a bursting poof at ends of twigs. **FRUIT:** Pea pod, but instead of flat, *constricted between seeds*, making it look like a string of pearls; begins green and matures to yellow.

SPECIES REMARKS: The green twigs, fireworks-like appearance while flowering, and pearl-like fruit set this species apart, along with the dark black buds enclosed by the petioles.

369

Black Locust
Robinia pseudoacacia

A very common tree across the Northwest, black locust is found often as a street tree but even more often as a self-seeded invader of roadsides and vacant or ignored landscapes. It can form thickets when left alone for too long and often falls apart due to weak branch attachments and unions.

Large tree up to 80 feet (24 meters) tall, with upright, layered, bunchy habit, reminiscent of a dry tropical savannah tree. **BARK:** Gray to brown; gets thick quickly, developing hard, intense, interweaving ridges, deep furrows. **TWIGS:** Often zigzag at nodes, reddish-brown, angular with regular lenticels; a *pair of stipular spines* is present at each node, sharp and roselike; buds very small, often several per node, like a clumpy mass of tissue with individual buds obscured. **LEAVES:** 6–14 inches (15–35 centimeters) long, odd-pinnate with 7–19 leaflets; leaflets 1–2 inches (2.5–5 centimeters) long, *oval, with rounded tips*, sometimes slightly or minutely pointed; slightly bluish-green, hairless when mature. **FLOWERS:** Perfect, white, fragrant, about 1 inch (2.5 centimeters) across, showy; borne densely on long, pendulous racemes in early spring. **FRUIT:** Flattened pod, green maturing to dark brown; not more than 4 inches (10 centimeters) long, numerous, often persistent through winter.

SPECIES REMARKS: Black locust is often confused with honeylocust and pagodatree. The small, flattened pods; thick, dark bark; and stipular spines set it apart from both, which have very different fruit and bark.

Honeylocust
Gleditsia triacanthos

This native of eastern North America is frequently planted as a street and park tree in our region, as it's able to grow in conditions that other trees would find unfavorable, ranging from wet bioswales to dry, compacted tree wells. It's common throughout the region but isn't invasive.

Large tree up to 70 feet (21 meters) tall, with upright, vase-shaped habit; predominantly single stemmed; has comparably thin, airy canopy that casts dappled shade. **BARK:** Gray, platy when young, becoming more fissured with age as hard, woody plates reflex back from sides; often with intense, frighteningly long, branched thorns. **TWIGS:** Green to yellowish-brown, often zigzag at nodes, which are swollen; older twigs have knobs from which leaves and flowers sprout and which tend to accumulate moss and lichen; buds small, yellow, dome-like, sometimes a few at each node. **LEAVES:** 2 kinds: newest growth tends to have mostly *bipinnately compound leaves*, with each pinna further split into 8–14 leaflets, while older twigs have clusters of primarily pinnately compound leaves with 20 or more leaflets; often no terminal leaflet (even-pinnate); leaflets small, usually around 1 inch (2.5 centimeters) long or less, lime green to yellowish; rachis and rachilla pubescent. **FLOWERS:** White, fairly inconspicuous; not useful for identification. **FRUIT:** Long (8–18 inches, or 20–45 centimeters), flattened pod about 1 inch (2.5 centimeters) wide or wider, the biggest of all legumes in our region; flat, but often becoming curly when drying.

SPECIES REMARKS: If you see showy flowers that catch your attention, it's not a honeylocust. But if you see giant pea pods, knobby twigs, and a mix of pinnate and bipinnately compound leaves, then you know you've got one.

LEFT: *Pinnate leaves* **MIDDLE:** *Bipinnate leaves*

Silk Tree
Albizia julibrissin

Probably one of our more exotic trees, the silk tree (also called mimosa) seems to come straight out of a tropical savannah. It's short-lived in our region, but it's planted as an ornamental in yards and along streets for its unique and vibrant flowers. While it's not invasive here, in the southeastern United States it's a problem.

Small tree up to 30 feet (9 meters) tall, with spreading, often low canopy; usually quickly splitting into a few main scaffold limbs; airy canopy. **BARK:** Mostly very smooth, gray, developing a long split every now and then. **TWIGS:** Green to gray-brown, with many prominent lenticels; buds small, rounded, often several stacked at one node above large leaf scar. **LEAVES:** *Bipinnately compound*, up to 20 inches (50 centimeters) long in some cases, but often smaller, with 10 or more pinnae and dozens of leaf-lets; each leaflet uniformly sized and arranged, usually around ⅜ inch (1 centimeter) long, densely packed; matte green; late to leaf out. **FLOWERS:** Terminal clusters of pink-topped poofs, like the hair of a troll doll; technically flowers are tiny and the showy parts are long pink stamens; fragrant, very showy in late summer. **FRUIT:** Flattened pea pod, 5–7 inches (13–18 centimeters) long, green maturing to light brown; often present along with flowers.

SPECIES REMARKS: The bipinnately compound leaves with fine, delicate leaflets and the pom-pom flowers easily set silk tree apart. Comparatively, honeylocust's leaves are not as finely bipinnate, and its pods are much larger.

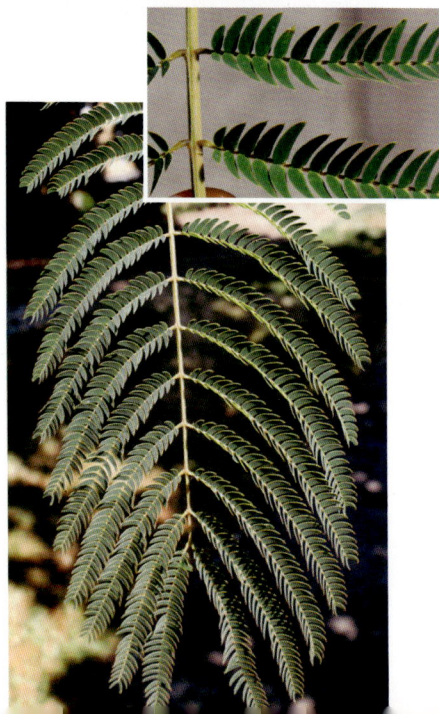

Kentucky Coffeetree
Gymnocladus dioicus

Kentucky coffeetree is being planted more and more in our region. They can be quite fickle to get started from seed, but once established, they are very tough and not invasive. Found in parks and as street trees, they are not quite a popular yard tree yet.

Medium tree up to 50 feet (15 meters) tall, with upright, rounded or somewhat irregular crown on single stem; twigs large, well spaced; winter silhouette can be striking. **BARK:** Similar to honeylocust's, but the scaly plates are smaller, creating a wavy, curvy pattern of slightly uplifted woody plates. **TWIGS:** Stout, olive green to reddish or gray, with lenticels; usually hairless, but sometimes slightly downy; buds brown, set into the stems, not projecting far above surface; a few small buds can poke out above large, wide, U-shaped leaf scars, with large vascular bundle scars present. **LEAVES:** *Bipinnately compound, huge*, not uncommonly over 3 feet (1 meter) long with 3–7 pairs of pinnae, each with 6–14 leaflets, usually offset along rachilla; each leaflet ovate with pointed tip and entire margin. **FLOWERS:** Dioecious mostly; male flowers small, hardly noticeable on short panicles; female flowers more noticeable on larger panicles, more fragrant; neither will jump out at you. **FRUIT:** Flat yet thick and succulent pod, very leathery, about 5–10 inches (13–25 centimeters) long, slightly curved; begins light yellow-green, then fades to dark brown; can remain on tree in winter; filled with large, disc-shaped seeds.

SPECIES REMARKS: The size of the bipinnately compound leaves and unique, pulpy fruit will set Kentucky coffeetree apart. In winter, look for the unique bark, any lingering fruit, and sunken buds.

Goldenchain Tree

Laburnum × watereri

Barely making the cut, this small tree is a hybrid of the only two species in the genus, both native to Europe. It's found mostly as a bedraggled street tree or yard tree, and every now and then it's found in natural areas due to its invasive tendencies.

Small tree, grows to 15 feet (5 meters), rarely taller; upright with a single stem; messy, half-dead canopy with basal sprouts. **BARK:** Greenish, even on larger stems, turning brown with age and developing diamond-shaped splits. **TWIGS:** Olive green to gray, usually hairless, but popular cultivar 'Vossii' has pubescence; buds short, ovoid, covered in fine, silky hairs; often on short spur shoots. **LEAVES:** *Trifoliate*, up to 3 inches (8 centimeters) long; matte green above and below, hairless at maturity; leaflets almost sessile. **FLOWERS:** Long, pendulous racemes of bright yellow flowers appearing in early summer, often more than 1 foot (30 centimeters) long. **FRUIT:** Small, flattened pods that hang on racemes, often persisting through winter.

SPECIES REMARKS: The rarer parent species of goldenchain tree are Scotch laburnum (*L. alpinum*) and common laburnum (*L. anagyroides*); all three are difficult to tell apart from one another. Scotch laburnum has glabrous twigs, nearly glabrous leaves, and blooms weeks earlier than common, which has pubescent twigs and leaves. Like other hybrids, goldenchain tree often has a mixture of traits from both parents.

Chinese Pistache
Pistacia chinensis

This tree is gaining in popularity as a street tree on the west side of the Cascades, as it's a tough tree with an agreeable habit. Sadly, it's not the species the delicious nuts come from, but it's ornamentally quite nice. It's not invasive, so it's found only as a street or landscape tree. **Family: Anacardiaceae**

Medium tree, usually 30 feet (9 meters) tall, with round, fairly tight canopy that reminds me of Raywood ash. **BARK:** Gray to light brown, breaking into small, scaly pieces further broken up by shallow, orangish fissures; can become slightly exfoliating with age. **TWIGS:** Stout, dark to light brown, with prominent lenticels; newest growth often downy, gray; buds dark brown to red, rounded, slightly pointed away from twig. **LEAVES:** *Even-pinnate, lacking a terminal leaflet;* up to 10 inches (25 centimeters) long;

each leaflet 2–4 inches (5–10 centimeters) long, oppositely paired, with entire margins. **FLOWERS:** Dioecious; male flowers in dense clusters in early spring, female flowers in loose clusters, both on last year's growth, so new leaves come out beyond flowers. **FRUIT:** Small, pea-sized drupe held in great masses on large panicles; mature in fall to either red or blue.

SPECIES REMARKS: The even-pinnate leaves are the first clue to look for if you suspect this species; almost all others will have odd-pinnate leaves. The red and blue fruit in the fall also sets it apart, along with the striking bark. You may need to look closely at the buds in winter to be sure of the identity.

375

THE LAST OF THE ROSE FAMILY

Mountain-ash (family Rosaceae) are common in many gardens in the Northwest. A few shrubby native species can be found in the wild forests, but the most common tree in cities and towns is European mountain-ash (also called rowan). Being in the rose family, these trees are not closely related to true ash trees or to the eucalyptus species that are commonly called mountain-ash.

The genus *Sorbus* has recently been dissected, and many of the species have been split out into their own genera; whitebeam (*Aria edulis*, previously *Sorbus aria*) and Korean mountain-ash (*Alniaria alnifolia*, previously *Sorbus alnifolia*) are two examples (see p. 288). The most up-to-date names are included here, but many resources still list all these species under *Sorbus*, so make sure to check the scientific names closely when you're out exploring to avoid confusion (if that's even possible).

All trees in *Sorbus* have flat-topped clusters of flowers similar to the hawthorns, to whom they are closely related. However, they lack thorns and generally have far more flowers per cluster.

European mountain-ash flowers, twig, and fruit

European Mountain-Ash
Sorbus aucuparia

Also called rowan, this species is a favorite for its large puffs of flowers in spring and very colorful leaves and fruit clusters in fall. However, it's unfortunately short-lived, invasive in some of our areas, and prone to breakage, so it's not as loved as it once was. Look for it along streets and in yards.

Medium tree no more than 40 feet (12 meters) tall, with upright habit, becoming slightly more rounded with age (if it doesn't break apart). **BARK:** Gray, smooth, often shiny; can develop small pockmarks, but generally doesn't get too rough or cracked. **TWIGS:** Stout, reddish-brown, slightly pubescent at first, then more glabrous; buds large, pubescent, conical, but slightly curved so point is not directly in the middle of the bud. **LEAVES:** Odd-pinnate, 5–9 inches (13–23 centimeters) long, with 9–15 leaflets, each usually around 1–2 inches (2.5–5 centimeters) long; leaflets lanceolate with acute tip and serrations along margin, but usually none on bottom third. **FLOWERS:** Perfect, white, small, but in large corymbs nearly 5 inches (13 centimeters) across; do not smell nice. **FRUIT:** Small, berrylike pomes, several in clusters in late summer, starting out green, then maturing to orange or red.

SPECIES REMARKS: The serrated, odd-pinnate leaves will set it apart from many other species that have pinnately compound leaves with entire margins. The flowers and fruit can be confused with those of hawthorns, but they are larger and more numerous, and hawthorns don't have such nicely pinnate leaves. **American mountain-ash** (*S. americana*) may look similar, but its leaves are slightly longer, at 6–12 inches (15–30 centimeters), with more leaflets (eleven to seventeen), which are usually intensely serrated down to their base (rather than only the top two-thirds).

377

Golden Raintree
Koelreuteria paniculata

Common in parks, yards, and mostly along streets, golden raintree (also called Chinese lantern tree) can be found in nearly any part of the Northwest. It's not considered invasive here, but it has been known to seed itself in elsewhere, so it is one to watch.
Family: Sapindaceae

Medium tree, up to 35 feet (11 meters) tall, with rounded, somewhat spreading, serpentine crown; very uniform density in wintertime. **BARK:** Light brown to grayish-brown, rough; breaks up into vertically oriented, somewhat crossing ridges; very uniform appearance. **TWIGS:** Stout, orangish-red to light brown, with prominent lenticels and no hairs; buds dark brown, triangular to pyramidal, with distinct point and 2 obvious valvate scales; leaf scar triangular, raised. **LEAVES:** Curiously pinnate to bipinnate, up to 18 inches (45 centimeters) long; leaflets near base and tip are small with incised lobes (each sort of like an individual hawthorn leaf); in the middle, they get bigger, more lobed, often transitioning to being entirely bipinnate. **FLOWERS:** Perfect, yellow, borne in large, loose panicles that cover the tree like fireworks; very showy. **FRUIT:** Dry, papery, 3-part capsule not more than 2 inches (5 centimeters) long, containing a few small, black, pea-sized seeds.

SPECIES REMARKS: The fruit of golden raintree tends to stay on through winter, but the unique leaves, buds, and flowers will help set it apart year-round. **Staghorn sumac** (*Rhus typhina*) is almost never tree sized, so it only gets an honorable mention here. It has pinnately compound leaves similar to golden raintree's, but it's much smaller overall, with very different fruit.

PALMS

Arecaceae

Palms are mostly denizens of tropical and subtropical climes, and they're botanically unique compared to other trees covered in this guide. Palms are **monocots** (a shortened version of *monocotyledons*), a group of flowering plants that includes grasses, lilies, Joshua trees, dragon trees, and many others. These plants all start their lives by sending up a single embryonic leaf called a **cotyledon** from the seed, hence their name meaning "one cotyledon." All the other broadleaf trees in this guide send up two cotyledons from the seed, thus they are called *dicotyledons*, or **dicots**. (Conifers also have cotyledons, but in varying amounts.)

Their unique qualities don't end there. Palms have just a single growing tip called the heart (technically a **meristem**, akin to a single, giant terminal bud). It's responsible for upward growth and is the only part of a palm that produces new leaves (called **fronds**), flowers, and fruit. However, it does not produce branches, and the stem does not increase in diameter as it gets taller because palms do not produce wood in annual rings like other trees. Taken together, this unique anatomy gives palms their characteristic form. (Incidentally, this also gives some people leave to declare that palms are in fact not trees, as they lack several characteristics that a traditional "tree" has, such as branching and woody annual growth, but I'll leave that debate for another time.)

The Pacific Northwest has just a single representative from the palms common enough to be included here. A few other palms and palmlike, treelike plants can be found too, especially along the southern Oregon coast, though they're rare. Refer to *A Californian's Guide to the Trees Among Us* by Matt Ritter for more on palms.

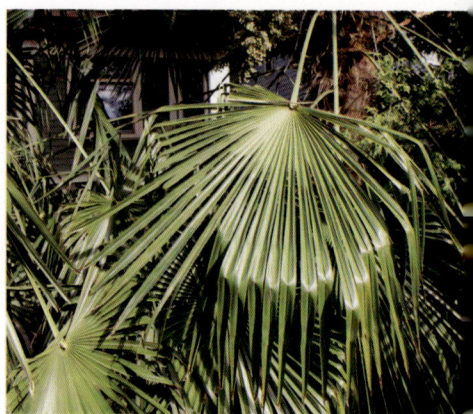

Chinese Windmill Palm
Trachycarpus fortunei

As the only real palm that we have growing around here, I almost don't need to tell you any more about it: if you see a healthy palm growing in our region outside of the southern Oregon coast, it's assuredly a Chinese windmill palm, which some call Chusan palm. Find it growing as a landscape tree west of the Cascades.

Small tree up to 30 feet (9 meters) tall, with single stem; crown not very wide, topped with several large fronds; does not produce more than one stem; usually not more than 1 foot (30 centimeters) in diameter, which does not increase. **BARK:** Fibrously hairy, with old frond bases sticking out. **TWIGS:** Well, there ain't any; the only growing tip is the giant heart at the top of the stem. **LEAVES:** Huge, titularly palmately lobed; circular leaf over 3 feet (1 meter) in diameter, splitting into several long, thin, straight-veined, leathery lobes with sharply pointed tips; pleated at base where lobes have not split apart. **FLOWERS:** Tiny, yellow, borne in pendulous bunches from upper canopy, not wildly showy, but obvious when you see them. **FRUIT:** Many rounded, black, berrylike drupes covered in whitish bloom; hang pendulously from upper crown, often obscured by large leaves. **SPECIES REMARKS:** You've probably seen this tree and wondered, "What the heck is a palm tree doing here?" It's one of the only species in the family that is hardy enough to withstand our cool winters. You may come across a few other smaller palms, as well as some yucca (*Yucca* spp.), but none will reach tree height. You may see more palms in the future as more hardy species are tested in the region or our climate continues to change.

ACKNOWLEDGMENTS

This guide is surely the most ambitious project I've taken on, and I could never have done it alone. To start, countless people before me have studied these trees and written about them, and I am the beneficiary of their work and knowledge. Whether it's the peoples who have lived in our region and managed the landscape since time immemorial, or the people who sorted through the introduced species later on, I am indebted to their keen observations and efforts to keep track of things.

More personally, I need to first thank the educators, dendrologists, and arborists who first taught me the names of the trees around me and how they work. These include Whitey Lueck, Ann Bettman, Ed Jensen, and Paul Ries, some of my original professors and mentors at the University of Oregon and Oregon State University—Whitey being the first and most notable influence on the trajectory of my life with his natural history course Trees Across Oregon. Scott Baker, one of my first and most constant professional mentors, first encouraged me to teach a tree identification course, which led me onto the path I am on today.

While I was writing, my good friends and colleagues Rayna Gleason, Mandy Tu and Martin Nicholson (both with Hoyt Arboretum), Phyllis Reynolds, Preston Pew of Cistus Nursery, Brandon Namm, Ryan Gilpin, Tobin Mitnick, Michael Andrews with the City of Boise, Matt Perkins, and Nolan Rundquist with the City of Seattle were indispensable in providing guidance, encouragement, information, ideas, companionship, and even a few beers along the way.

I relied heavily on a few online resources and books, whose authors so kindly replied to my very esoteric questions when I reached out. These include Chris Earle with The Gymnosperm Database, Michael Kauffmann with Backcountry Press, and Patrick Breen with the Oregon State University's Landscape Plants Database.

Keith Turner and Kara Monroe of the University of Alaska at Anchorage, Taha Ebrahimi, and Dan Blanchard were instrumental in providing help and photos for some trees I simply couldn't get to, and I am so thankful for their efforts and excitement to be involved.

Special thanks must go to my dearest friends Dan Gleason, Sean Hogan of Cistus Nursery, Thomas Van Hevelingen, and August Schwartz. Each of these people has played an outsized role in my life as a friend and influence and has helped me over the finish line, whether by way of advice, expert opinion, or as a sounding board.

I'd also like to thank all my dear friends and family who have suffered me talking their ears off about trees from the first time I learned a scientific name to this very day. There are far too many of you to name, but suffice it to say if we've chatted about trees or you've ever asked me to identify a tree for you, here's looking at you. My community is why I am here doing this, and it truly means the world to me that I can share my passion with you all personally.

In particular, I'd like to recognize Carey Jacquinot and Alex Crowson for their friendship and support. Carey's encouragement, spirit, support, companionship, and willingness to listen to me talk about trees far too much kept me going; she has been a rock from start to finish during this project and beyond. Alex, my cohost, coproducer, and cocreator of *Completely Arbortrary*, but most importantly one of my closest friends and my creative other half, supported me in taking this on from its inception, both in spirit and with guidance on photography, art, and presentation. Without him and the community of people who have joined and supported us, this book would never have been written. With that, I should also thank all the Fungal Associates out there, that network of people across the world who have lifted me up and helped make this work possible. To anyone who has listened to me talk about trees or encouraged me, thank you.

Finally, thank you, Emily White, Laura Shauger, Jen Grable, and everyone else at Mountaineers Books, for taking a risk on me and this work. I hope I have not let you down.

GLOSSARY

abaxial Referring to a *leaf* or *cone scale*, the outer or lower side that faces away from the *axis*, or away from the sun. Synonym: dorsal.

accessory Of *buds*, set directly next to another bud, namely often the *terminal bud*.

achene A dry, *indehiscent fruit* that contains one seed, usually very small and appearing like a seed itself.

acorn The *fruit* of an oak; a hard *nut* (*calybium*) with a cap (*cupule*) covering the top.

acuminate A gradual taper to a pointed tip with concave sides.

acute A gradual taper to a pointed tip with straight sides.

adaxial Referring to a *leaf* or *cone scale*, the inner or upper side that faces toward the *axis* or toward the sun. Synonym: ventral.

alternate arrangement Having one *bud* or *leaf* at a given *node*, either set spirally along the *twig*, on alternating sides, or some degree of both.

angiosperms A group of vascular plants that produce *flowers* and whose *ovules* are held within an *ovary*; the mature ovary is the *fruit*, which contains the *seeds*. These are known as broadleaf trees.

anther The *pollen*-producing part at the end of the *stamen* held up by the *filament*.

apex Of a *leaf*, the tip farthest from the *petiole*.

apical meristem The growing tip on a *shoot*, often specifically referring to the bud at the end of the dominant main *leader*.

apophysis The tip of a pine *cone scale* that is exposed when it is closed. Plural: apophyses.

appressed Lying tightly against, as in a *bud* or *leaf* lying tightly against a *twig*.

arborescent Per the introduction (p. 10), *see tree*.

archetypal form The classic or typical form a given *species* takes, usually in an *open-grown* situation.

aril A fleshy covering over a *seed*; namely, the red fleshy bit around the seed of a yew (*Taxus*).

arrangement Referring to *leaves* or *buds*, their position relative to each other along a *twig*. See *alternate, opposite, sub-opposite,* and *whorled arrangements*.

asymmetric Referring to a *leaf base*, when the angle of attachment of the *blade* is significantly different on either side of the *petiole* but the *margins* connect at essentially the same point.

awl-like A *conifer leaf* type that looks like an awl; the leaf is often *appressed* to the *twig* at the *base*, then reflexes away from it while tapering to a sharp point.

axil The area between two diverging parts in reference to the *axis*, such as between a *twig* and the *leaf petiole* that diverges from it.

axillary Borne in the *axil*. Of *buds*, borne in the *leaf* axil; generally a synonym for a *lateral* bud.

axis A central structure from which parts emerge.

bark The outermost layer of tissue covering the *stems* of trees; produced by the cork *cambium*.

basal Referring to the *base* of a part, such as a main *stem* or *leaf*.

base Of a *leaf*, the section of the *blade* closest to where the *petiole* and blade connect.

berry A fleshy *fruit* with multiple *seeds* inside.

bilateral symmetry Referring to a *flower*, the property of having the left and right sides as mirror images of each other, such that an imaginary line drawn vertically would produce two symmetric halves, but a line drawn in any other orientation would not.

binomial A two-part *scientific name* made up of a *genus* and *specific epithet*; a *species* name.

bipinnate Of a pinnately compound leaf (*pinnate*), having *pinnae* that are further divided into pinnately compound *leaflets*.

bisexual flower One that contains both male and female parts. Synonym: *perfect flower*.

blade The flattened, expanded body of the *leaf*; the *lamina*.

bloom Of *flowers*, one that is fully open. Otherwise, a waxy coating that covers plant surfaces, usually *leaves*, *fruit*, and *twigs*; it's white or light gray and easily rubs off. *Stomatal bloom* refers to the bloom found on leaves, often on the underside. A tree part that is covered with bloom is called *glaucous*.

bract A modified *leaf*; often contributes to a *flower* display but is not strictly a part of the flower or the *fruit* that follows. *See also scale bracts.*

branch architecture The general pattern or shape of the collective *stems*, *limbs*, and *twigs* of a tree's *canopy*, including, but not limited to, characteristics like texture, angularity, density, size, and branch orientation.

broadleaf A general term for *flowering trees*. Contrasted with *conifer*. Synonym: *angiosperm*.

bud A bundle of embryonic tissues (*shoots*, *leaves*, and/or *flowers*) set in summer for the next growing season; often (but not always) covered by *bud scales*.

bud arrangement The number of *buds* per *node* along a *twig* and their position relative to one another. *See alternate, opposite, subopposite,* and *whorled arrangements.*

bud scale scars Scars left by the *bud scales* that encircle the *twig* after the scales fall away in spring. These generally mark the successive years of growth along the twig: from the tip backward, each set of scars marks one year's growth.

bud scales The coverings enclosing the embryonic tissues of a *bud*; technically, highly modified *leaves* or *stipules*.

calybium The *nut* of trees in the beech family (Fagaceae); paired with a *cupule*.

calyx The collective term for the *sepals* of a *flower*.

cambium The vascular system of a *tree*, just below the *bark* and just outside the *wood*.

canopy The upper portion of a *tree* that bears the *leaves*; the *crown*.

capsule A dry, *dehiscent fruit* that splits open to release its *seeds*.

catkin A usually pendulous *inflorescence* composed of many *unisexual flowers* tightly packed together along a central *axis*. Each flower has a subtending *bract*, and they are usually not showy. Also the term for the mature collection of *fruit* that comes from the female flower catkins of some species, notably those in the birch family (Betulaceae).

cauliflory A flowering trait characterized by *flowers* and subsequent *fruit* emerging from larger woody *stems* and *limbs*; not associated with injury or *epicormic sprouts*.

ciliate Fringed with fine hairs.

cladode A modified *stem* that performs photosynthesis; unique in our region to umbrella-pine (*Sciadopitys verticillata*).

codominant leaders Two or more main *stems* that grow upward competing for apical dominance and do not appear to significantly diverge from each other as in *scaffold limbs*.

columnar Having an upright growth *habit* that is very tightly spaced.

common name The vernacular name of a *tree* used in common parlance.

complete flower One that has a *calyx*, *corolla*, *pistils*, and *stamens*.

compound Referring to a *leaf*, broken up into two or more discrete sections where the *leaf blade* is not connected as a single unit.

cone The reproductive structures of *conifers*, consisting of *scales* and *bracts*; of two kinds: *seed cones* (female) and *pollen cones* (male).

conifer A plant that bears a *cone*; *ovules* are not enclosed in an *ovary*.

cordate Heart shaped, with two large, rounded lobes on either side of a *midvein* or *petiole*; can refer to a whole *leaf* that is heart shaped or only to the *base* of a leaf with two rounded lobes.

coriaceous Leathery, often in reference to a *leaf* or *fruit*.

corolla The collective term for the *petals* of a *flower*. (Also a very dependable car.)

corymb An *inflorescence* with a flat or rounded top where the individual *flowers* arise from different points along the *peduncle*, and the lowest flower stalks (*pedicels*) grow longer and outward to be roughly level with the shorter upper ones; blooms from the outside in.

cotyledon An embryonic *leaf* that is first to emerge from a *seed*; in *angiosperms* there is either one or two cotyledons, while *conifers* can have several. *See also dicot* and *monocot*.

crenate Having rounded teeth along the *margin*, like so many semicircles. Noun: crenation.

crenulate Having *crenations* that are very small, usually referring to a *leaf margin*.

crown General term for a tree's *canopy*.

cultivar Portmanteau of the words "cultivated" and "variety," meaning a *variety* of a plant that has been artificially bred or cultivated for a specific set of traits.

cuneate Wedge shaped, referring to a symmetrical *leaf base* with straight sides that tapers as it goes down the *petiole*.

cupule The cap or sheath-like covering on *fruits* in the beech family (Fagaceae) consisting of a cupped structure that partially or fully encloses the *nut* (called a *calybium*).

cutleaf A *leaf* type (usually on a *cultivar*) characterized by abnormally deep sinuses between lateral veins and serrated margins, giving the leaf a lacy appearance. Synonym: *laceleaf*.

cyme A flattened *inflorescence* where the central *flowers* bloom first.

deciduous Loses its *leaves* perennially; can be *broadleaf* or *conifer*.

decurrent Having a general growth *form* characterized by an upward-spreading *habit* that develops from branches that diverge and grow away from each other; classically develops a rounded shape.

decussate Having a growth *habit* (usually referring to *twigs*, *leaves*, or *buds*) characterized by parts in opposite pairs being oriented 90 degrees relative to each other, creating an X or + pattern when viewed from the end.

dehiscent Referring to *fruit*, one that splits open or apart at maturity; referring to a *seed cone*, one that breaks apart at maturity, scale by scale. Noun: dehiscence. Antonym: *indehiscent*.

deltoid Triangular (referring to the Greek letter delta, Δ), with the *petiole* at the base of the triangle.

dentate Toothed, with the teeth pointing more outward than forward, like so many triangles, along the *margin*. Noun: dentation.

denticulate Having *dentations* that are very small, usually referring to a *leaf margin*.

dicot Short for "dicotyledon," it's one of the two major historical divisions of plants, characterized by *seeds* that produce two embryonic leaves (*cotyledons*) when first germinating.

dimorphic Occurring in two forms; often referring to *leaf* or *twig* growth where two distinct and consistent forms are apparent.

dioecious A *tree* that produces only unisexual flowers: either *pistillate flowers* or *staminate flowers* are produced on a given individual.

dormant Not physiologically active, usually during the winter in the Pacific Northwest. Of *buds*, usually referring to *latent* buds that are not physiologically active until triggered hormonally.

drupe A fleshy *fruit* that has its *seed* inside a hard inner covering (endocarp), like a cherry (*Prunus*).

elliptical Narrowly oval shaped; thickest in the middle and equally skinnier at the ends; when broadly elliptical, it's just called *oval*.

emarginate Notched at the *apex*, as in a *leaf* or petal.

entire Referring to a leaf's *margin*, having no distinct texture or divisions; smooth.

epicormic sprout A vegetative *sprout* that appears from main *stems* or *scaffold limbs* in response to stress, damage, or a change in light conditions.

epidermis The outermost layer of tissue on a plant, including *stems*, *leaves*, *fruit*, and *flowers*. On older stems it is replaced by the *bark* with age.

espalier A *tree* trained in a geometric pattern and in one plane, often against a wall.

even-pinnate Referring to a pinnately *compound leaf* (*pinnate*), one that has an even number of *pinnae*, and often lacking a *terminal leaflet*.

evergreen Keeps its *leaves* for more than one season.

excurrent Having a general growth *form* characterized by a strong central vertical leader and *limbs* that radiate outward mostly horizontally; classically develops a triangular shape.

exserted Extending beyond or sticking out.

extreme heterozygosity A genetic trait of some *species* characterized by the offspring (produced by sexual reproduction) of two individuals developing traits that are unpredictable and often radically different from its parents'.

facial Of *scale-like leaves*, those that are on the top and bottom of the *twig*.

falcate Referring to a *leaf*, flattened and curved laterally, like the blade on a sickle.

family The *taxon* level above *genus* that broadly includes groups with similar *flower* or *cone morphology*.

fascicle A bundle of *needle-like* leaves on a pine (*Pinus*); each fascicle usually contains two, three, or five needles.

fastigiate Having an upright growth *habit*; often a *cultivar* type.

filament The "stem" of the *stamen* that holds up the *anther*.

fissures Referring to *bark*, long, wide grooves, usually vertically oriented, that usually give the bark the appearance that it's cracking open or ripping apart.

floral Referring to *flowers*; of buds, containing only embryonic flower tissues.

flowers The sexual reproductive organs of a plant that may consist of *pistils*, *stamens*, *petals* (*corolla*), *sepals* (*calyx*), *tepals*, and/or *ovaries*.

foliage Generally, the leaves of a tree.

foliar Relating or in reference to the *leaves*.

forest grown A growth *form* of *tree* within a grove or closed-canopy forest; characterized by a single skinny *stem* with *limbs* and *foliage* present only very high in the canopy, at approximately the same height as the other trees around it.

form The general appearance of a tree's *canopy*, focused specifically on its shape and the direction or orientation of its growth (upright, spreading, compact, pendulous, rounded, etc.); can be affected by its surrounding environment (i.e., *open grown* away from other trees or *forest grown* with many trees around it).

fronds The large *leaves* of palms; also the term for fern leaves.

fruit A mature *ovary* containing *seeds*.

furrows Referring to *bark*, long, deep grooves, usually V shaped, in thick bark, often with pronounced ridges between them.

genus The *taxon* level above *species* and below *family*, indicating close kinship;

designated by similar *flower* and *fruit* morphology. Plural: genera.

glabrous Lacking hairs.

glaucous Covered in a light-colored, waxy coating that can be rubbed off, as on blueberries or plums.

globose Globe-like, spherical.

graft A physical (artificial) joining of two genetically distinct plants, consisting of a *scion* (*twig*) or a *bud* that is attached to a *stock*; commonly used to propagate desired *species* or *cultivars* by taking cuttings of them and attaching them to new rootstock. Can also happen naturally between roots of the same species or branches on one tree.

gymnosperm A group of vascular plants whose *ovules* are not held within an *ovary*; *conifers* are the primary representatives of this group in our region.

habit The specific way that a *species* tends to grow, irrespective of its environment.

hard pine General term for pines (*Pinus*) with needles in *fascicles* of two or three.

husk An outer coating, usually on a *fruit*; tends to be ephemeral, decaying away or drying out.

hybrid The offspring of two distinct *species*, usually displaying a combination or mixture of the parent trees' traits; can be naturally occurring or artificially created.

hypanthium A cup-shaped *flower* base formed by the fusing of the *calyx* and/or *corolla* at the base of a flower; commonly seen on flowers in the rose family (Rosaceae).

imbricate Of *bud scales*, overlapping one another, usually with several scales on one bud.

incised Of *margins*, deeply and sharply cut margins, usually irregularly.

indehiscent Referring to *fruit*, one that does not split open at maturity; referring to a *seed cone*, one that does not break apart at maturity. Antonym: *dehiscent*.

inflorescence A collection of *flowers*, including their stems (*peduncles* and *pedicels*) and any *bracts*.

infructescence A collection of individual *fruits* that are not strictly fused together but generally appear as one unit.

invasive species A *species* that does not naturally occur in our region (i.e., it arrived here after European colonization) and can easily reproduce, creating self-perpetuating populations apart from human cultivation.

involucre A collection of *bracts* or *leaves* that grow at the base of a *flower* or *inflorescence* or a *fruit* or *infructescence*.

laceleaf A *leaf* type (usually on a *cultivar*) characterized by abnormally deep *sinuses* between lateral veins and serrated *margins*, giving the leaf a lacy appearance. Synonym: *cutleaf*.

lamina The *leaf blade*.

lanceolate Lance shaped; skinny and much longer than it is wide, usually referring to a *leaf*.

latent Of *buds*, hidden; set surreptitiously below *bark* and remaining *dormant* indefinitely. When activated, often in response to outside stimuli, it will usually create new *shoots*, but it can become new roots under certain circumstances.

lateral Of *buds*, set along the length of the *twigs*; not *terminal*. Of *scale-like* leaves, those on the left and right of the twig.

leader A main *stem* that has *apical* dominance and tends to grow upward; most other *limbs* arise from it or grow away from it.

leaf The organ that arises from below a *bud* and usually performs photosynthesis; may be *simple* or *compound*. Plural: leaves.

leaf scar The mark left on the *twig* by the *petiole* once the *leaf* falls away, always just below the *bud*.

leaflet A discrete section of a *compound leaf* whose *blade* does not directly connect with the blade of another leaflet; there is no *bud* at the base of a leaflet.

legume A *dehiscent fruit* that splits along two suture lines; a "pea pod"; definitional fruit type of plants in the pea family (Fabaceae).

lenticel A small, often raised, corky bump on *twigs* and *stems* that facilitates gas exchange.

limb A large branch.

lobe An extended part of a *leaf* that is not separated from the rest of the *blade*; like a peninsula of leaf blade; separated by *sinuses*.

lobed Referring to a *leaf*, having distinct *lobes* that do not break into *leaflets*; often described as *pinnate* or *palmate*.

lumper A cheeky term for taxonomists who tend to lump *taxa* together under fewer *species* or *subspecies* based on broader traits; assumes more intraspecific variation.

margin Of a *leaf*, the outermost edge of the leaf *blade*. Referring to a landscape, the edges that are often less managed or ignored, near boundaries, borders, or changes in ownership of some kind.

meristem Undifferentiated stem cells in plants that divide and create new tissue, namely located in the tips of roots and *shoots*, as well as the *cambium*.

midvein The central vein in a *leaf* or *leaflet* from which other veins diverge, usually in a *pinnate* arrangement; the obvious extension of the *petiole* through the leaf *blade*.

monocot Short for "monocotyledon"; one of the two major historical divisions of plants, characterized by *seeds* that produce one embryonic leaf (*cotyledon*) when first germinating.

monoecious Having both *staminate flowers* and *pistillate flowers* borne on the same tree; flowers are not *perfect*.

monotypic Of a *genus* that has just one representative species (e.g., *Umbellularia*); can also refer more broadly to any *taxon* (i.e., one genus in a *family*).

morphology The physical traits of a *tree* or a specific part or organ; broadly, how something grows.

mucronate Abruptly pointed, often referring to the end of a *cone scale* or the *apex* of a *leaf* or *leaflet*.

naked Of *buds*, no *bud scales* present; embryonic *leaf* and/or *flower* tissues are exposed throughout the *dormant* season.

native species A *species* that grows naturally in our region and grew here prior to European colonization.

needle-like A *leaf* type characterized as being long and skinny, much longer than it is wide, and generally not tightly *appressed* to the *twig* for any of its length; can be *singly borne*, in bundles (*fascicles*), or in clusters; often referred to simply as needles.

node A point along the *twig* where a *bud* and/or a *leaf* is borne.

nonnative species A *species* that does not naturally occur in our region; distinct from an *invasive species* in that it does not necessarily create a self-perpetuating population.

nut A dry, *indehiscent fruit* with a single *seed* and a hard outer shell; some have *husks* that fully or partially cover them.

nutlet A very small *nut*, often appearing as a *seed* but botanically is a whole *fruit* with a tiny seed inside.

oblique Referring to a *leaf base*, the *blade margin* does not connect to the *petiole* at the same point on either side but instead is offset (e.g., elms).

oblong Two to four times longer than it is wide, generally rounded.

obovate Egg shaped, but the *apex* is wider than the *base*.

obtuse Referring to a *leaf*, rounded, not pointed, almost circular.

odd-pinnate Referring to a pinnately compound leaf (*pinnate*), having an odd number of *pinnae*, usually made up of paired *leaflets* along the *rachis* and a single *terminal* leaflet.

open grown A growth *form* of a single tree or a small group of trees grown apart from others, characterized by a large, symmetrical *canopy* with large *limbs* and *foliage* equally distributed on all sides and from very low along the main *stem*; often has a thick main stem and/or several main stems or large *scaffold limbs* beginning low in the canopy.

opposite arrangement Having two *buds* and/or *leaves* on either side of a *twig* at a given *node*.

orbicular Circular, or very nearly so.

orchard tree One planted and managed specifically to produce *fruit*.

order The taxonomic rank above family that includes closely related families and their constituent genera and species.

ornamental species Usually a *nonnative species* (but not necessarily) that is intentionally planted and maintained for an aesthetic purpose.

oval Shaped like an oval: *oblong* yet symmetrical; broadly *elliptical*.

ovary The chamber at the base of the *pistil* that contains the *ovule* or ovules in *flowers*.

ovate Egg shaped: rounded with a *base* that is wider than the *apex*, often in reference to a *bud* or a *leaf*.

ovule Female gamete enclosed within the *ovary* of a *flower* or set on the *scales* of a *cone*; the precursor of a *seed*.

palmate Like the palm of your hand, with multiple *axes* radiating from a central point at the end of the *petiole*; referring to a *compound leaf* or a venation pattern.

panicle A compound *raceme*, on which each branch branches again.

pedicel The stalk of one *flower* in a cluster (*inflorescence*).

peduncle The stalk of a solitary *flower* or of a collection of flowers before they split into *pedicels*.

peltate Shield-like, usually referring to the *scales* of many cypress family (Cupressaceae) *cones*; characterized by a scale that grows outward from a central *axis* and whose end is flattened perpendicularly to the rest of the scale, like a shield being held in an outstretched arm.

pendulous Distinctly hanging down; weeping.

perfect flower One that has *stamens* as well as one or more *pistils*. Synonym: *bisexual flower*.

perfoliate Referring to opposite leaves (*see opposite arrangement*), when their *bases* are conjoined and the leaf encircles the *node* (for example, *Eucalyptus* spp.)

petal One of a group of appendages in a *flower*, often showy, that surround the *stamens* and/or *pistils*; can be almost any color; collectively called the *corolla*.

petiole The stalk of a *leaf*.

petiolule The stalk of a *leaflet* on a *compound leaf*.

pinna Individual *leaflets* of a *compound leaf*. Plural: pinnae.

pinnate Feather-like with a central *axis* that branches on either side, referring to a *compound leaf* or a venation pattern.

pistil The female reproductive part in a *flower*, consisting of the *stigma*, *style*, and *ovary*.

pistillate flower A *unisexual flower* that bears only *pistils* and produces *fruit* from a mature ovary; a female flower.

pith The central area within a *twig* or *stem*.

plant awareness disparity Previously termed "plant blindness," a phenomenon where people do not notice the plants around them; PAD.

pollen Minuscule powdery grains that contain the male gametes and are produced in the *anthers* on the *stamen* of a *flower* or in the male *cones* of a *conifer*.

pollen cone The male *cones* that produce *pollen*; often small, papery, and short-lived.

pome A fleshy *fruit* with a compound *ovary* (with multiple *seeds* inside); for instance, an apple or pear.

prickle A sharpened protuberance from the *epidermis* or *bark*, usually derived from a *trichome*; not derived from a modified *leaf* (*spine*) or *stem* (*thorn*).

pseudo-terminal Of *buds*, appearing as the *terminal* bud, but is in fact the most recently set lateral bud, with a visible leaf scar and often small dead *twig* tissue beyond it.

pubescent Generally hairy. Noun: pubescence.

raceme An *inflorescence* similar to a *spike*, but each flower has its own *pedicel* attached to a central *axis*; blooms from the bottom up.

rachilla The main *axis* of a *pinnae* on a *bipinnately compound leaf*.

rachis The main *axis* of a *compound leaf*.

radial symmetry Referring to a *flower*, it's the property of having all the parts radiate out from essentially a central point such that no matter how it is rotated, the pattern of the parts is consistent and repeats itself; multiple imaginary lines could be drawn through the central point in different orientations and each would produce two symmetric halves.

radicle The embryonic root that emerges from a *seed* opposite the *cotyledons*.

ranks Referring to *leaf* orientation, more or less discrete rows of leaves that grow out in distinct orientations, often left or right of the *twig* (*two-ranked*) or in three distinct planes (*ternate*) or in four different directions (*decussate*); referring to *taxonomy*, a level among the hierarchy.

receptacle The base of a *flower* that connects it to the *peduncle* or *pedicel*.

reiteration A vertical *leader* (usually one of multiple) that often grows and develops in response to the loss of the initial dominant, apical leader, but can occur from side limbs that simply turn upright for some reason.

revolute Referring to the *margin* of a *leaf*, with an upper edge that curls slightly below the leaf.

rounded Not straight or flat; convex.

rugose Wrinkled.

samara A dry, *indehiscent fruit* type characterized by a single *seed* with a usually papery wing growing off or around it (e.g., ashes, elms). When two samaras are connected at their base they are called a double samara; this is the definitional fruit type of the maples (*Acer*).

scabrous Rough to the touch, sandpapery. Synonym: *scurfy*.

scaffold limbs Large *limbs* attached to a main *stem* that hold up large portions of the *canopy* and are dominant architectural components of the *crown*.

scale A general term referring to a flattened, *appressed* part or appendage, often covering a surface such as a *stem* (in reference to *bark*), a *leaf*, a *twig*, or *fruit*. Referring to cones, the dry, woody structures that bear the *ovules* and later the fertilized seeds on *seed cones* (and technically the *pollen* on *pollen cones*), usually growing out radially from a central *axis*; they are often either flattened

and spirally arranged (as in pine family *cones*) or *peltate* and in opposite pairs (as in many cypress family cones). *See also bud scales.*

scale bracts Of cones, appendages that grow along with the *scales* of many *conifer* species, such as noble fir (*Abies procera*) and Douglas-fir (*Pseudotsuga menziesii*).

scale-like A *leaf* type characterized as being tightly *appressed* to the *twig*, often in pairs or threes, such that the leaves are nearly indistinguishable from the twigs.

scientific name The official *binomial* name of a plant, written in Latin, and all its constituent parts, including *subspecies*, *variety*, *cultivar*, etc. It is botanically important, as it defines where a tree lies among the *taxonomy* of life.

scion The upper part of a *graft*, usually the desired variety; placed on top of the *stock*.

scurfy A description of a surface (usually a *leaf* or *twig*) that is rough to the touch; sandpapery. Synonym: *scabrous*.

seed The reproductive package held within a *fruit*; contains fertilized embryonic plant cells that can germinate into a new individual.

seed cone The female *cones* consisting of several *scales* that each hold one to several *ovules* that develop into *seeds* once fertilized; larger and woodier than pollen cones.

sepals The outermost group of appendages around a *flower*, often what forms the flower *bud*; usually green, but often showy in their own right; collectively called the *calyx*.

serrate Toothed, with the teeth pointing toward the *apex* of the *leaf*. When serrations have serrations of their own, they are doubly serrate.

serrulate Having *serrations* that are very small, usually referring to the *margin* of a *leaf*.

sessile Lacking a *petiole*, *peduncle*, or *pedicel*.

sheath Tissue that wraps around the *basal* section of the *needles* in a *fascicle*; falls away on five-needle pines, but is usually retained on two-needle and three-needle pines.

shoot The newest growth arising from a *bud*, including the new *stem* tissue, *leaves*, and/or *flowers*; often in reference to the earliest growth in spring, not yet hardened into a woody *twig*.

silhouette The outline of a tree showing its overall *form*, *habit*, and *branch architecture*, often during the winter.

simple Referring to a *leaf*, one with a single *blade* area that is connected together with no breaks; includes leaves with deep *sinuses* as long as the *lobes* are still connected.

single scaled Of *buds*, having one *scale* that envelops the bud and falls away in one piece.

singly borne Referring to parts that do not grow in tight clusters but appear as discrete individuals, namely *leaves* or *flowers*.

sinuate Wavy, often describing a surface or a growth *habit*, as a *stem* or branch (as opposed to a *margin* or edge, which is called *undulate*); muscly looking. Synonym: sinuous.

sinus The space between *lobes* on a *leaf* or *leaflet*.

soft pine A general term for pines with *needles* in *fascicles* of five.

species The primary unit of *taxonomy*; a discrete, natural population of individuals that share common characteristics and interbreed among themselves to create similar offspring; generally do not interbreed with other species; written as a *binomial* in italics, *genus* first followed by the *specific epithet* (e.g., *Abies procera*).

specific epithet The second half of a *binomial* or *scientific name*, following the *genus* and indicating the *species*.

spike An unbranched *inflorescence* with *sessile flowers* along it that bloom from the bottom up.

spine A sharpened protuberance usually derived from a modified *leaf* or leaf part.

splitter A cheeky term for taxonomists who tend to split up populations into many discrete *taxa*; assumes less intraspecific variation.

spread The width of a tree's *canopy*.

sprouts New *shoots* arising from *latent buds*, often in response to stress or damage; generally don't follow the natural pattern of growth displayed throughout the rest of the *tree's canopy* and are less consistent with normal *species* traits.

spur shoot A short secondary *shoot* along a *twig* (the primary shoot); does not significantly elongate, so *buds* set along it are very tightly packed, appearing *whorled*.

stamen The male reproductive part in a *flower*, consisting of the *filament*, which holds up the *anther*. Plural: stamens.

staminate flower A *unisexual flower* that bears only *stamens* and produces only *pollen*; a male flower.

stem Broadly, the woody, aboveground parts of *trees* that produce *bark* and to which *buds*, *leaves*, *flowers*, and *fruit* attach; specifically, a main trunk of a tree that generally comes from the ground.

stigma The tip of a *pistil* that receives *pollen*.

stipule A leaflike appendage at the base of the *petiole*.

stock The lower part of a *graft* to which the *scion* is attached; often a hardier or more generic *species*.

stomata Pores on *leaf* surfaces that allow for gas exchange; often associated with distinct patterns of waxy *bloom* called *stomatal bloom*. Synonym: stomates. Singular: stoma, stomate.

stomatal bloom A waxy, whitish substance associated with the *stomata* of a *leaf*.

style The "stem" of the *pistil* that holds up the *stigma* and connects it to the *ovary*.

subfamily A taxonomic division of a *family* that indicates a closer relationship between its members; above the genus level.

subopposite arrangement Having *buds* and/or *leaves* set in consistent pairs that are slightly offset from each other; often they are fully *opposite* elsewhere along the *twig*.

subspecies A naturally occurring taxonomic rank below *species* and above *variety*; usually defined as a subpopulation with unique characteristics that is geographically separated from other subpopulations.

suckers New *shoots* that sprout from *latent buds*, often referring to those growing at the base of a *stem*.

synonym A defunct *scientific name* for a *species*, resulting from some taxonomic update. Abbreviation: syn.

taxon A discrete unit of *taxonomy* that is separable from others by some defined characteristic(s), such as a *species*, *variety*, *subspecies*, *hybrid*, *genus*, or *family*. Plural: taxa.

taxonomy The art and science of naming and categorizing life into hierarchical groups based on shared characteristics.

tepals *Flower* appendages that are not differentiated into *petals* and *sepals*; used mostly to describe magnolia flowers, which have no discernable differences between petals and sepals.

terminal Referring to an *inflorescence* or *fruit*, borne on the end of a *shoot* and representing the end of the shoot growth. Of *buds*, set at the end of the *twig*, thereby ceasing the stem elongation. Referring to a *leaflet*, the last leaflet on a *pinnately* (*pinnate*) *compound leaf* representing the leaf's *apex*.

ternate Having parts in distinct groups (*ranks*) of three.

thorn A strong, sharpened protuberance derived from a modified *stem;* can be branched or singular and can have *buds* or *leaves* attached.

tomentose Densely hairy.

trade name A marketing name given to a *cultivar*; not necessarily the official cultivar name, but rather one meant to promote sales of the plant.

tree Per the introduction (see p. 10), *see arborescent*.

trichome A tiny, often single-celled growth from the *epidermis*; superficially, a tiny hair.

trifoliate Describing a *compound leaf* that splits into three distinct *leaflets*.

truncate Squared, often referring to the *base* or *apex* of a *leaf*.

twig The most recent one to three years of growth of a lateral branch.

two-ranked A *foliage* orientation characterized by two distinct rows of *leaves* usually growing opposite each other along a *twig* on a single plane.

type genera A *genus* that defines its *family* or *subfamily*.

type species A *species* that defines its *genus*.

umbel A flattened or rounded *inflorescence* with the *pedicels* all emerging from one point; can be upright or *pendulous*.

umbo A projection at the tip of a *scale* on a *cone*.

undulate Wavy, often referring to a *leaf margin* or *canopy* trait.

unisexual flower One that bears only male (*staminate*) or female (*pistillate*) parts.

valvate Of *bud scales*, two scales enclosing the *bud* that do not overlap, like a clamshell.

variety A naturally occurring taxonomic rank below a *species* or *subspecies*; usually defined by some characteristic(s) that is unique to a subdivision of a species. The subdivision will freely interbreed with others in its species and is not necessarily geographically separated from them.

vascular bundles Vascular tissues within a *petiole* that connect the *leaf* to the vascular system of the *tree*; most obvious as small dots in the *leaf scars*.

vegetative Of *buds*, containing embryonic *leaf* and *stem* tissue, and sometimes also *floral* tissue.

water sprout A *sprout* that emerges from the base of the *stem* or seemingly randomly throughout a *canopy* with no apparent damage or stress; a *sucker*.

whorled arrangement Having three or more *buds* set at a single *node*.

RESOURCES

General

Farjon, Aljos. *A Natural History of Conifers.* Timber, 2008.

Hanson, Thor. *The Triumph of Seeds.* Basic Books, 2015.

Jonnes, Jill. *Urban Forests: A Natural History of Trees and People in the American Cityscape.* Penguin, 2017.

Mitnick, Tobin. *Must Love Trees: An Unconventional Guide.* Rock Point, 2023.

Peattie, Donald Culross. *A Natural History of North American Trees.* Collins, 2007.

Pollet, Cédric. *Bark: An Intimate Look at the World's Trees.* Frances Lincoln, 2010.

Tudge, Colin. *The Tree.* Three Rivers Press, 2005.

Yoon, Carol Kaesuk. *Naming Nature: The Clash Between Instinct and Science.* WW Norton, 2010.

Native Trees

Arno, Stephen F., and Ramona P. Hammerly. *Northwest Trees: Identifying and Understanding the Region's Native Trees.* Mountaineers Books, 2020.

DeMarco, Lois, and Jay Mengel. *Identifying Trees of the West.* Stackpole Books, 2015.

Farrar, John Laird. *Trees of the Northern United States and Canada.* Wiley-Blackwell, 1991. (Also available as *Trees in Canada.*)

Jensen, Edward C. *Trees to Know in Oregon and Washington.* 70th anniversary edition. Oregon State University Extension Service, 2021.

Kauffmann, Michael Edward. *Conifers of the Pacific Slope.* Backcountry Press, 2013.

Lanner, Ronald M. *Conifers of California.* Cachuma Press, 1999.

Parish, Roberta, Ray Coupé, and Dennis Lloyd. *Plants of the Inland Northwest and Southern Interior British Columbia.* Lone Pine Publishing, 2018.

Pojar, Jim, and Andy MacKinnon. *Plants of the Pacific Northwest Coast: Washington, Oregon, British Columbia & Alaska.* Revised edition. Lone Pine Publishing, 2016.

Ritter, Matt, and Michael Kauffmann. *California Trees: A Field Guide to the Native Species.* Backcountry Press, 2025.

Rowe, Carol A., Robert W. Lichvar, and Paul G. Wolf. "How Many Tree Species of Birch Are in Alaska? Implications for Wetland Designations." *Frontiers in Plant Science* 11 (2020): article 750. https://doi.org/10.3389/fpls.2020.00750.

Sibley, David Allen. *The Sibley Guide to Trees.* Knopf, 2009.

Turner, Mark, and Ellen Kuhlmann. *Trees and Shrubs of the Pacific Northwest.* Timber Press, 2014.

Ornamental and Landscape Trees

Dirr, Michael A. *Manual of Woody Landscape Plants.* Sixth edition. Stipes Publishing, 2009.

Dirr, Michael A., and Keith S. Warren. *The Tree Book: Superior Selections for Landscapes, Streetscapes, and Gardens.* Timber Press, 2019.

Ebrahimi, Taha. *Street Trees of Seattle: An Illustrated Walking Guide.* Sasquatch Books, 2024.

Jacobson, Arthur Lee. *Trees of Seattle.* Second edition. Author, 2006.

Price, Elizabeth A. *Native and Ornamental Conifers in the Pacific Northwest: Identification, Botany, and Natural History.* Oregon State University Press, 2022.

Reynolds, Phyllis C. *Trees of Greater Portland.* Timber Press, 1993.

Ritter, Matt. *A Californian's Guide to the Trees Among Us.* Expanded edition. Heyday, 2022.

Online Resources

American Chestnut Foundation: tacf.org

Calscape, California Native Plant Society: calscape.org

Flora of North America: floranorthamerica.org

The Gymnosperm Database, edited by Christopher J. Earle: www.conifers.org

The New York Botanical Garden: www.nybg.org

Oregon State University's Landscape Plants Database, curated and managed by Patrick Breen: landscapeplants.oregonstate.edu

Trees and Shrubs Online, maintained by the International Dendrology Society: www.treesandshrubsonline.org

The USDA Plants Database: plants.usda.gov

Virginia Tech Dendrology Factsheets: dendro.cnre.vt.edu/dendrology /factsheets.cfm

INDEX

ABOUT THE AUTHOR

Casey Clapp is an arborist, dendrologist, and educator based in Portland, Oregon, and the cocreator, coproducer, and cohost of the science podcast *Completely Arbortrary*. An energetic and enthusiastic storyteller, Casey leads people through the world of trees (our world) via a lens of science, history, and real-world cause and effect. He holds advanced degrees in forestry and environmental conservation and is recognized as a Board Certified Master Arborist by the International Society of Arboriculture.

When not looking at trees . . . well, he's probably sleeping.